CLARE,
A SINGLE SENTENCE.

A SINGLE SWALLOW

Running for the Hills

Truant: Notes from the Slippery Slope

Sicily: Through Writers' Eyes

A SINGLE SWALLOW

Following an Epic Journey from South Africa to South Wales

HORATIO CLARE

Chatto & Windus
LONDON

Published by Chatto & Windus 2009

2 4 6 8 10 9 7 5 3 1

First published in Great Britain in 2009 by
Chatto & Windus
Random House, 20 Vauxhall Bridge Road,
London SW1V 2SA
www.rbooks.co.uk

Addresses for companies within The Random House Group Limited can be found at:
www.randomhouse.co.uk/offices.htm

The Random House Group Limited Reg. No. 954009

598.
826

A CIP catalogue record for this book
is available from the British Library

Hardback ISBN 9780701183127
Trade Paperback ISBN 9780701183134

The Random House Group Limited supports The Forest Stewardship
Council (FSC), the leading international forest certification organisation. All our titles
that are printed on Greenpeace approved FSC certified paper carry the FSC logo. Our
paper procurement policy can be found at www.rbooks.co.uk/environment

Mixed Sources
Product group from well-managed
forests and other controlled sources
www.fsc.org Cert no. TT-COC-2139
FSC © 1996 Forest Stewardship Council

Typeset by SX Composing DTP, Rayleigh, Essex
Printed and bound in Great Britain by
Clays Ltd, St Ives PLC

Maps by Jeff Edwards

Dedicated to the memory of Deon Glover

Contents

Preface

Some years ago I sat on the tarmac at Bole airport in Addis Ababa, Ethiopia, talking to a soldier. We had been chatting about not much for a little while when his officer approached. This man's skin was very black and his eyes were hot, red and suspicious. He shouted at the soldier in Amharic, and then he barked at me:

'Where are you from?'

'The UK,' I said, 'I am British.'

I was squinting up at him against the light. My face was sunburned.

'No,' he said, angrily, as if his suspicions had been confirmed. He jabbed a finger at me.

'You are Russian,' he said, scornfully.

I stuttered. The officer was right, in a way. My father's mother was a White Russian who fled the revolution, with her mother. She met my grandfather in Shanghai. Over half a century later Russian soldiers, 'advisers', had come to Ethiopia to assist the Marxist dictator, Mengistu: this officer must have recognised their fair hair, their sunburned skin and Slavic features in me. As I struggled to pacify the officer I felt a deep blush, as much in me as on my skin. Nothing I could say would erase my face or the story it told him. It was the only time I have felt the fear and humiliation so many have known, of being condemned for the tribe you belong to, of being unable to escape or disguise your ethnicity and the perceived sins of your people.

Yet, for all the bitterness this man felt against the Russians with whom I shared an ancestry, and for all his evident suspicion of what I had been doing – talking to his soldier, spying, perhaps, inveigling myself into the confidence of the farmer's boy who I had discovered behind the dusty uniform and the hefted gun – I was doing nothing more subversive, as I sat on the tarmac, waiting for a plane, than watching birds.

'There are no frontiers in the sky,' my mother was fond of saying, quoting T. H. White, who gives the words to a goose in *The Sword in the Stone*. I was brought up to believe that all people were equal. Opposition to prejudice and discrimination was a family tradition, I told myself, proudly, when I was young: my father had been banned from South Africa, where he grew up, for opposing the Apartheid regime. As a child of the Cold War I covertly supported the Russians, when I learned I was quarter Russian, and though I had never been to South Africa, I believed – romanticising freely – that the blood of a South African freedom fighter ran also in my veins. My parents divorced when I was seven; seeing little of my father made him all the more heroic to me.

An imagined Africa was part of the landscape of my childhood. We lived high up on a sheep farm in South Wales; we had a hill at our back, rearing over the house, and a view across the valley to a range of mountains which carved the far skyline into waves, cuts and crests.

'It looks just like Africa,' my mother said (she had been to Kenya), and then she quoted Karen Blixen: ' "I had a farm in Africa, at the foot of the Ngong Hills . . ." '

And we had visitors from Africa, too. They nested in the Big Barn and perched on the telephone wire; they filled the whole summer with their comings and goings and their twittering calls.

'Look! The swallows are back!' we said, when they arrived, and 'Ah, the swallows are going,' my mother announced, with a kind of mourning, in the autumn, when they gathered on the wires.

She marvelled at the great distances the birds would travel, and wondered at what they would see, on their journey to Africa, and we bade them farewell, and hoped that they would return safely.

When we moved down to the foot of the mountain the house we bought had an attic with an unsafe floor, broken windows and swallows' nests on its beams.

'It was one of the reasons I chose it,' my mother said.

She often said she would like to be a swallow when she died, which I understood: in the back of your mind, in a dreamy way, you could not help but want to go with them. In your hand a swallow weighs little more than a full fountain pen, yet twice every year it makes a journey of a scale and precision unmatched by our mightiest machines. Theirs is an extraordinary existence, even by the standards of birds.

The Cold War ended, and apartheid was swept away. I no longer 'support the Russians', and have grown out of claiming my father's deeds as my own. Although our valley is little changed, the farm I grew up in is no longer a ramshackle smallholding but a comfortable home for a retired couple. The buildings have been modernised: only the Big Barn is still as it was, and every year the swallows still come and go. But though the lives of the children of our valley are now more pressured and less isolated than those of my brother and I, at root the expectations of the young, and what is expected of them, are perhaps much as they were. If we do not stay on our farms, and eventualy take them over from our parents, then we are supposed to go out into the world, make our fortunes – or at least our own way – find someone to love, and put down roots. By the age of thirty-three I had not honoured this tradition. I had not settled down with anyone, nor had children. I had no home of my own.

One bright morning in the late summer of last year, I opened an upstairs window and startled five swallows on a wire. They were only a couple of feet away and taken completely by surprise. In the instant before they leapt into the air I saw them in vivid detail: the red patches, like the masks worn by medieval knights, covering their faces and chests; their dark blue backs, creamy underparts and the marbled span between their long tail feathers. That was all it was, and I did not know it, but everything that followed began then.

Within a couple of days of seeing the birds I had a plan, a scheme for an adventure and an education. In January the swallows would be in their wintering grounds, thousands of miles away in the heat of the South African summer: I resolved to fly to South Africa, find the birds and follow them all the way home. I would trace the flight of migrating birds between the South Africa my father came from and the Welsh hillside where my mother raised me. I would test the beliefs that I had been taught: that people, regardless of creeds and colours, are equal; I would rely on the best of the oldest of humankind's traditions: kindness to strangers. By devoting myself to one thing, these birds, I fantasised that I might discover – or at least understand better, in microcosm – something of the working of the world. And beyond this, by following swallows, I hoped to put down a marker on my own life, something to separate the boy-man I was from the man I wanted to become.

I spared no expense on preparation: everything I had I threw at visas, vaccinations and equipment for the expedition. At the same time I embarked on a crash-course of research into swallows. It soon became clear that I had had no idea of how extraordinary were the creatures I blithely planned to follow, or how powerful, old and strange are the relationships between us and them. As they gathered on the wires, ready for their departure, I stared at them, and wondered what they were saying.

CHAPTER I

South Africa: Travelling Companions

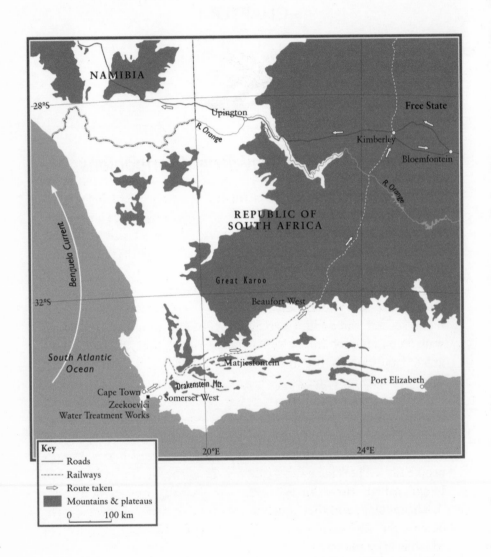

South Africa: Travelling Companions

I am Inkonjani, the lightning bird, the breaker of cliffs; I am the bird that brings the rain. My name is Nyankalema, the one who never gets tired; I am Tififiliste, I am Ifilelis, you have heard me say my names. I am Giri Giri, a magical charm: if you could catch and eat me you would be protected from car smashes, plane crashes, boating accidents and train wrecks, proofed against all the perils of the road – but only for five years!

For I am the snapper-up, the here-and-not-here. Through me Isis crossed from life to death to visit her murdered beloved, Osiris. (One night several thousand years ago I fluttered around his remains, walled up in his column in Byblos.) I am sacred, holy, a sign of God's grace. I put out the fire in the Temple in Jerusalem – have you not seen me scoop up water as I go?

My name is Sisampema (in Kwangali) and Lefokotsane, and Malinkama, as they know me in South Soto. Swael is my Afrikaaner name. The Shona people say 'Nyanganyena!' when they see me (you have nearly heard that before), like 'Nyenga', which the Tsonga people say. (The Tsonga also call me Mbwalana – in many languages I am called two things.) The Xhosa call out 'Udilhashe!', 'Ulelizapholo!', and they must be very fond of me, or know me well, because they also know me as the Lightning Bird and say 'Inkonjane!', which is my Zulu name.

Here in the south of Africa I am *intaka zomzi*, a bird of the home. None of the Nguni peoples would ever harm me, because I bear a message from their ancestors. Life in the next world is richer, better: none who honour me should despair.

'The Zulus say,' my father said, thoughtfully, when I told him my plan, 'that those who follow the swallows never come back.'

That frightened me. I began to think of ways around it: perhaps the Zulus mean that if you follow them from what is now Kwa-Zulu Natal you will not go back to Kwa-Zulu Natal. *Well that is all right, I am not going to Kwa-Zulu Natal*, I told myself. But I could not help thinking about it.

Somewhere on a cape of Africa, in a tropic of half-imagined time, where proud people lived in kraals, ruled by leaders great and terrible, Mzilikaze, Chaka, who led them to the wars in impis, whose footsteps shook the earth itself, one day someone, perhaps a young man, told his family he was going to follow the Inkonjani and went away with them, just before the summer ended and the storms of winter came. Saying farewell to his friends and passing on foot from village to village, whatever trail he left of memories and encounters, whatever snatches of news came back to his family, if he had one, or to his friends, if they lived, nothing of him survives but a certainty in a saying. Those who follow the swallows . . . He cannot have been the only one.

Apart from very rare exceptions all the swallows of the north flee south before the winter. Born in the high latitudes in early summer, they fly through an autumn which heats and brightens to a sea, then becomes a desert. The south-bound route for British birds runs in a rough line from Brighton to Barcelona to Cape Town, cutting corners around the Gulf of Guinea. (If the map of Africa is an elephant they go straight across its ear and slide down its trunk.) They fly over plains of permanent sun to forests of constant rain; across glades and hills and wooded worlds of mist, through the valleys of rivers, some copper, some green, to the grasslands, the vast savannahs towards the end of the land, and the veldts and the vleis of southern Africa. Here they base themselves near lakes, marshes and rivers, where they roost.

Arriving in November at the beginning of the southern summer, the swallows remain in the south until just before the onset of winter in February and early March. They spend their time here eating, travelling a bit and roosting in large flocks. They court a little, and compete perhaps for food, though the southern African summer does not lack flies. They certainly do not prospect for nest sites, mate, build, lay, incubate, rear young or defend territory: all this they save for the north.

To follow the birds from south to north I constructed a route plan entirely based on two sentences, from the best and most recent book on swallows from a British perspective, *The Barn Swallow* by Angela Turner.

The return migration in spring is more direct [than the southerly autumn route]. There are then more records from Algeria and Tunisia than in the autumn and more from the central Mediterranean between the Balearics and Italy. British Barn Swallows move north through Europe along the eastern coast of Spain and the western coast of France . . .

My route, based on the paragraph above, should therefore link South Africa, Algeria and Wales, bearing north-north-west.

The swallow pages of *The Migration Atlas* of British birds include a map showing Africa and Europe, and a curvaceous, red, many-tailed arrow joining the Republic of South Africa to the United Kingdom of Great Britain and Northern Ireland. This is a computer-plot, made from joining the dots of ring recoveries. (Trapping swallows, ringing them and recording where those rings are recovered, either on birds trapped for a second time, or found dead, is still the only way in which ornithologists are able to track their movements.) I stared at it, imagining a vast wind-tunnel of swallows hurtling diagonally up the globe.

Our two species, swallows and humans, have lived alongside each other throughout recorded time. The earliest recorded instance of swallows and men living together is a 15,000-year-old nest in a cave in

Derbyshire – a cave which was also inhabited by man. Amazingly, we who live and work by information superhighways do not know very much more about swallows than did Gilbert White in 1789, when he recorded the then common speculation that they spent the winters clinging together in the mud at the bottoms of European rivers and ponds. They were supposed to join claw to claw, beak to beak, and slide down the reeds in which they roosted. In his dictionary Dr Johnson defined 'Conglobulate' as hanging together in clumps, as swallows were supposed to do. The stories that gave rise to this belief remain a puzzle. Two hundred years before White, Olaus Magnus wrote:

> when winter cometh [. . .] in the northern waters fishermen oftentimes by chance draw up in their nets an abundance of Swallows, hanging together like a conglomerated mass . . .

There are stories from Germany which go further, telling of birds recovered in this manner which came out of 'hibernation' in the warmth of a fisherman's house and flew around the room before expiring.

How can this be explained? A kind of cryogenic lightning, like the icy winter of 1682–3, which froze the Thames for two months? Perhaps speculation is half the pleasure of the amateur birdwatcher; if so, we must thank professional ornithology for the other half.

The swallow and martin family, *Hirundindae*, occur almost everywhere on earth except for the polar caps. In the northern summer the Barn Swallow might be found anywhere in Eurasia, from the furthest west of Ireland to the Central Himalayas, and has a North American cousin, very little different, which breeds from Hudson Bay to Baja California and migrates to Mexico. They all follow the same seasonal tides, coming and going, breeding in the north. There are a great many kinds of swallow along the migration route of the Barn Swallow: Mosque Swallows, Angolan Swallows, Greater Striped Swallows, Red-Rumped Swallows, Pearl-Breasted Swallows, not to mention the swifts and martins. Some of these creatures travel half the world twice a year, others do not migrate, but merely shuffle around their preferred parts of the globe.

Swallows do not yet submit to satellite tracking: they are too small and fly too far for the technology of the time of writing. Ornithologists do not yet have a transmitter which would last the time the migration takes, and still allow the birds to fly – the batteries are too big. The only way to follow them, therefore, is to go with them.

But they are fast, so fast: 4 metres per second in low gear, 14 metres per second at top speed. Depending on wind, route and inclination they can cover 300 kilometres in a day. They can do the entire trip in twenty-seven days – perhaps less. My plan was to leave South Africa with the vanguard of the migration, and to arrive with the main body, or even with the tail-enders. In this way I hoped to travel with the migration, rather than trying to keep pace with individual birds. My timetable was built on a departure from South Africa at the beginning of February and an arrival in Britain around the middle of April, when most birds should be making their landfalls.

Their summering grounds in South Africa have flowed and ebbed, perhaps with climates, perhaps with changing human land-use. From strongholds in the tropical eastern Cape they spread westwards, and may now be in retreat again. Ringing data suggested that the best place to find a bird that came from near my home in South Wales, in the west of western Europe, would be in the far south-west of the Cape.

While the route was vague, the birds' speed intimidating and the reputations of some of the countries to be crossed either obscure or outright ominous, one or two aspects of swallow behaviour gave me hope. They fly at low level, feeding as they go: between a foot above the ground and 60 feet up is typical. So if there were any around, I ought to be able to spot them. And no less than the height of flight, the habitats they favour are dictated by the whereabouts of their prey: blow-flies, hover-flies, beetles, aphids, moths, caterpillars and water bugs, among other insects. Watercourses, rivers and lakes would always be good places to look for them. They favour reed beds for roosting on migration, so swamps might prove fruitful. When temperatures are high they tend to hunt and fly at lower levels and further away from vegetation, nearer to it and higher up when the weather is cooler.

Swallows prefer to fly into or across the wind, which allows them enough lift to hunt as they go, feeding with no fear of stalling on whatever crosses their path. When they have to make time, crossing the Sahara, for instance, they may go high up and ride a tail wind. When they find somewhere conducive to rest and feeding they may pause for days at a time.

On 26 January 2008 I woke an hour before the light and lay still. For weeks I had marvelled at nothing much: at the turning of a tap that brought warm water, at the blissful plunge into a clean towel, at the smell and softness of everything, at the opulence and ease of life: at piles of food, at PIN numbers, at films on desire and all the rest.

Today was the end of all that. My belongings were in storage, my savings were liquidised: I had bet everything on today and thrown everything at it. In two bags I had everything I could possibly need. Still I knew I was not ready, not prepared enough, not well-read enough, not pared down enough, but now it was almost time.

I lay still and breathed through fears. Knifed in South Africa, infected in Zambia, cursed in Congo, battered in Cameroon, murdered in Nigeria, kidnapped in the Sahel, slaughtered like a sheep in Algeria and probably posted on the internet. Money gone, passport gone, phone gone, no way out . . .

Then the morning came, and lunch with my brother, and then afternoon. Just before leaving the house I sat on my rucksack and shoulder bag, in the hall. This is a Russian tradition: you sit quietly on your bags before you begin a journey. You still yourself, despite your haste, and when you are ready, you go. So I sat on my bags, half-thinking that I would at least remember anything I had overlooked. I had forgotten nothing. I was comprehensively, almost laughably, well equipped.

In my back pocket was a UK passport, number 108949308, in the name of David Horatio Clare, British Citizen, born 05 09 1973,

London. Inside the passport was one of travel's great paragraphs, the very definition of British self-image, which starts so grandly and becomes a hopeful sort of mutter:

> Her Britannic Majesty's
> Secretary of State
> requests and requires in the
> name of her Majesty
> all those whom it may concern to allow
> the bearer to pass freely without let or hindrance,
> and to afford the bearer such assistance
> and protection as may be necessary.

The passport contained a silicon chip, on a loop of what looked like copper wire, which as I understand it is a transmitter and receiver, holding unknown personal data. Between the chip and the request from Her Majesty's Secretary of State were the visas.

South Africa's would be granted on entry; Zambia's cost £30 and was valid for three months. There was a man asleep on his arms at the desk behind the receptionist in the Embassy of Cameroon. Cameroon cost a processing fee, a few days, many phone calls and a reservation confirmation fax from the Hilton, Douala. Congo-Brazzaville required endless phone calls (the lines are jammed during the day, apparently oil companies are responsible for much of the traffic) and cost US $150 sent via Western Union to the manager of the Hotel du Centre, Brazzaville, for the necessary fax confirming reservation, and a processing fee. Once it has established that you are not mistakenly seeking a visa for the Democratic Republic of Congo a man's voice gives painstaking directions to the consulate of the Republic of Congo, a broom cupboard in the corner of the offices of a commercial visa service in South London.

'Shall I bring my yellow fever certificate?'

'Well you can bring it but frankly I won't have time to look at it,' said the French accent on the phone.

'How long does it take to issue the visa?'

'I will issue it in about sixteen minutes,' he said.

Congo-Brazzaville's diplomatic mission to the United Kingdom consists of one man, Louis Muzzu, a thin, dark-haired Frenchman who would not look out of place in a photograph of de Gaulle's France, which, as a young man he served, staying on to work for post-independence regimes.

'It's a beautiful country,' he said, not exactly sighing, having stamped the passport with an attractive circular design incorporating the head of a Congolese woman and the motto 'Unity, Work, Progress'. It was valid for two weeks.

'Is it safe?'

'Oh yes, quite safe now.'

Obtaining a Niger visa meant faxes, calls and a processing fee, plus a trip to Paris, where the consulate was technically shut but where a kind woman issued the visa anyway, in return for my promise not to go to the north of the country.

'Too dangerous,' she said. 'Truly, much too dangerous.'

The Nigerian Consulate near Trafalgar Square contained a hundred people, with different colour tickets, some slumped, in demeanour apathetic, others in unhappy, restless lines: occasionally someone would make a despairing rush at the service windows, only to be repelled. Fortunately the rules had recently changed, there was no time to process my application and so I was spared trial by queuing system. I would try again in Africa.

Algeria required a reservation confirmation from a hotel in Algiers, which was wonderfully easy. But lying to, or rather being vague with, the consular authorities was not possible. If you are an independent tourist, you need an itinerary and a profession.

'Author.'

'Author?' said the visa man. 'Writer?'

It was a strangely warm and rainy January day in London. I was cold-sweating.

'Yes! – a writer of books.'

'Writer of what books? About what? Osama bin Laden?' (He grinned.)

'NO! My first was the story of . . .'

'You must make a list,' he said, with a half-smile. The clock was racing. Two minutes to get the form accepted, or the office would shut and my passport would miss visa day. The Algerian Consulate in London stamps visas on the 21st of the month. Miss visa day, miss visa, miss country.

'Do you have a piece of paper?' I cried, desperately, scrabbling for a pen.

In the end, with a grin, I was granted a visa valid for eight days.

'You can go anywhere!' cried the man. 'But you must not miss the exit date.' (Algeria is the tenth largest country in the world.)

In my front left pocket was a wallet, containing a credit card for 'emergencies', an expired National Union of Journalists' press card, a UK driving licence, UK National Insurance card and a debit card.

In the rucksack was a document folder containing vaccination certificates for Hepatitis A, Hep. B, Hep. C, Yellow Fever, Typhoid, Tetanus, Polio, Cholera, Meningitis, all 'boosted' as necessary, and Rabies – recording a course of three injections. I also had a 'fake' (i.e unstamped) Yellow Fever certificate, picked up at the travel clinic for use as a decoy.

'Are you going to be around dogs, buses, public transport, will you be travelling away from cities, will you be eating . . . uncooked food?' asked the nurse.

'Yes, yes – yes.'

'Right. Better have everything.'

There was a Comprehensive Travel Insurance certificate, promising immediate evacuation from anywhere, regardless of limitless cost, and my birth certificate, showing that it had taken my mother some months to get around to registering my birth in Hammersmith. Divided up and hidden in various places throughout the rucksack, was cash: US $1,000 and €1000.

I had a mosquito net and three kinds of spray – the vicious chemical stuff, some herbal stuff and something else. I had a treated net and sufficient tablets to get me through the malarial zone, roughly calculated to start somewhere in Namibia and finish somewhere in the Sahel. I had my hat (Indiana Jones style, naturally, purchased in St James's, London, as was his) and a new pair of binoculars. I had a

'Blackberry', allowing expensive and occasional access to email and web, also calls, text messages and GPS, which did not often work but could be thrilling, giving latitude and longitude and the degree of accuracy to which it calculated its position: normally about 8 metres.

I had books: Angela Turner, *The Barn Swallow*; *The Lonely Planet Guide to Namibia and South Africa*; and a collection of the prose of Seamus Heaney: *Finders Keepers*.

In a sponge bag were rehydration sachets, Savlon, TCP and an emergency dental kit (numbing clove oil, wadding, a mirror and a sort of soft-tipped probe); elsewhere in the rucksack were a First Aid kit, a compass, a beautiful head torch, enough maps of varying scale and relevance to get me arrested as a spy virtually anywhere between Windhoek and London, a journal with a reward offered in the front for its return (US $50), pens, pencils, lightweight clothes, boots and a rainproof jacket. I had visions of walking miles through tropical rain in the forests of the Congo. I was not worried about the downpours, but feared that the weight of the rucksack would have me floundering in the mud.

After a few moments of sitting I stood up, and went, discovering later that the sitting had popped open a tube of sunscreen in my wash-bag. (The Russians also say that life is not a walk across an open field.)

I took a taxi to Paddington Station. As we climbed up onto the Westway a cold sun was setting behind us in a sky streaked with pewter and hard blue. The towers of London looked stark and hard against a flinty dusk. As it sank the sun threw an orange flood of light out of the west. There, over 100 miles behind us, was the journey's end. It was 12,000 miles by the way I was going. From Heathrow I called three friends to say goodbye. From one, Samrine, a blessing came: *Bismillah al rahman al rahem.*

'Say it as you get on the plane,' she says. 'You'll be fine darlin', you know you will.'

You can see why the Greeks made sacrifices before setting out on a journey. The decision to travel any significant distance is a decision to put yourself in harm's way, to place your fate in the hands of the gods of winds, waves and the road. In Erice, near the port of Trapani in Sicily, sailors used to visit the temple of Aphrodite (in Punic times the temple of Astarte) and there make offerings to her, in the belief that the goddess controlled storms. The tradition has survived in that the Madonna of Trapani is recognised by the Vatican as the patron saint of sailors. Because I am fond of the place and the story, I say a quick prayer to her. I will go like a sailor; the roads will be my ships and the countries will be seas. My religious convictions – suspicions would be more accurate – are non-aligned. In need, distress or exultation I will worship God in any language. The major faiths condense the multiple spirits and deities the ancients perceived into single figures.

'*Bismallah al rahman al rahem*,' I mutter. In the name of Allah the almighty the all merciful . . .

I fall into a doze as we cross into Algerian airspace. Just before I drop off I see fires down there, rosy blooms of flame in the dark desert: oil wells. Two hours later we are still over Algeria. My head swims with muzzy forebodings: this is mad, it cannot be done, the swallows are just too quick, it is just too far, the plan is a joke . . .

I think of my father, somewhere in Cape Town and looking forward to tomorrow, to showing me around for a few days before he sees me off. He is there for a couple of weeks, researching his book about South African history, traced through its literature. Strange that the swallows should take me straight to the part of the beginning of my own story which is a mystery to me. South Africa, my father's story; somehow our family's story, which I have heard about but never seen.

'As you see,' my father says, 'South Africa is white!' Black waiters zig-zag between packed tables, frowning at the effort of carrying and distributing so much, much food. The tables are packed, and every customer is white.

I laugh a little, '*Je-sus* . . .'

'It is a pretty spot though, isn't it?'

Cape Town's waterfront is an orderly jumble of boats and quays, of

drilling rigs, restaurants, bars and day trips, chopped at by the South
Atlantic, with a wind, this lunchtime, coming from behind us, from the
other side of Table Mountain, the southeaster from the Indian Ocean.
The city has all the beauty of San Francisco; the luminosity of light,
which sharpens colours to the peak of their intensity; the fresh sea winds
which never abate; the different levels of streets and houses which seem
to applaud the prospect of the ocean; the deep blue shadows and, out of
the wind, the golden heat of the sun. The rich areas on the skirts of
Table Mountain have an American opulence about them. Near where
we are staying is a ranch with tall blue-gum trees and fine horses, ridden
and groomed to a gloss. In Camp's Bay a white woman who looks like a
model queues at a supermarket checkout wearing tiny scraps of
transparent white cotton. The black man who serves her does not know
which way to look. There are super-cars, Ferraris and Lamborghinis,
jockeying for position in the freeway traffic, which hurtles along with a
kind of recklessness. There is a recklessness in the wind and an
untameable ferocity in the cliffs and abysses of the mountain. There is
a tension in the air, as if all who live well live on borrowed time, and the
millions more who live hard are running out of patience. San Francisco
is a hundred times more at ease with the San Andreas fault than is Cape
Town with the human earthquake that has not yet come, which merely
rumbles, daily, in the crime round-ups of the newspapers. It is as
though Table Mountain is a volcano.

'As you see,' my father says, the next day, 'South Africa is black!'

Now we are driving through Khayelitsha, a suburb where over a
million people are living in tin shacks. We are the only whites on the
road. Cape Town is cut into quadrants; it is half chessboard, half
minefield. The residents speak in colour code: here live Blacks, here
Whites, here Coloureds. Wealth, poverty and danger are distributed
accordingly. All cities are segmented, but nowhere I have been are the
lines so sharp and hard, or the penalties for crossing them so dangerous. I give up counting the 'Warning! Twenty-four-hour Armed
Response!' signs, and the private security vehicles and personnel.

'We used to go down there and bring up lobsters, boil them in seawater and sell them on the side of the road,' my father smiles.

We are pausing on the way to Chapman's Peak; I am on the lookout for sharks and whales, grinning and baffled by the sun and the wind and the Cape's geography: while my father is entirely at home I cannot even grasp which way is north.

'You're a South African, Dad!' I blurt.

He is momentarily amazed, then starts laughing. He was a schoolboy and a student here before his flight in 1963, when his friends and their friends were being shot, arrested, tortured, tried, imprisoned or banned. He was banned for over twenty years (they would not let him back even when his father was dying) but has returned several times recently, like the other 'swallows', as some Cape Town residents call those among their friends who reappear between November and March, for the southern summer.

We went to a poetry reading where one of Dad's friends was performing. It was held in a bar called 'A Touch of Madness' in the Observatory district, which is supposed to be 'mixed'. There were two black customers among all the whites. Dad's friend read a poem about interrogations by the Security Police. The audience nodded, ruefully. An open-mike session followed: a young man read a long, chaotic, hip-hop/beatnik piece in the style of William Burroughs. The audience shifted in its seats. Then an Irishman stood and sang, plainsong, Patrick Kavanagh's 'On Raglan Road'. The audience stilled and listened.

> On Raglan road on an autumn day I saw her first and knew
> That her dark hair would weave a snare that I might one day rue.
> I saw the danger yet I walked along the enchanted way
> And I said let grief be a fallen leaf at the dawning of the day.
> On Grafton Street in November we tripped lightly along the ledge
> Of a deep ravine where can be seen the worth of passion's pledge,
> The Queen of Hearts still making tarts and I not making hay
> O I loved too much and by such by such is happiness thrown away . . .

There were thirty people in the room, thinking different thoughts of different lovers, but as the song ended there was a single note in their faces, a melancholy for something lost, or something that never was: whether it was enchantment, happiness or harmony you could not tell. I thought of her, again. The dream of all romantics, the dream you should perhaps have grown out of, by now; the one, the realised mystery, your own equivalent of Kavanagh's dark-haired girl, waiting for you, somewhere along the road of your life, with a spell of certainty attending her: the promise that the moment you see her, you know.

We went looking for the birds in a place called Zeekoevlei, 'Sea Cow Marsh', a little nature reserve between the Cape Flats and the sea. There was sun and wind and tall reed beds, there were hides for birdwatchers and high viewing platforms which we climbed because they were there; the wind made you hold tight to rails and posts. There were the tracks and droppings of hippos, there were ducks and stilts and all sorts of other pretty things I did not care about. There were no swallows.

'Seen any swallows recently?' had become Dad's catchphrase, but he was not saying it now. Where were they? I had not seen one since late last summer, in Britain. My appallingly expensive, near-unbeatable binoculars, light in their skeletal frame, their lenses hand-ground in eastern Germany, wobbled across crisp blue air.

'Try the sewage farm,' said one of the rangers. 'They come here later.'

Of course, it was mid-morning: insects; hunting. Reeds and roosting are for evening. We returned to the hire car and drove in circles, nosing the unobtrusive little vehicle through a strange, angular world of water, fenced-off settlements and tall trees until we are possibly lost, definitely lost, and then probably on track.

CAPE FLATS WATER TREATMENT WORKS

NO ENTRY

We play our joker again. We have done it before, using the car park and lavatories of the Victoria Hotel in town, because we were white or because our colour gave us confidence, though we did sneak out like thieves, and now, penetrating one set of gates after another because we are birdwatchers and birdwatchers can do this.

'There – swallow!' I shout.

'Is it?'

'Yes, look . . .'

'Oh so it is – phew.'

They were there for a second, skimming alongside us, between a high bank and the car. We follow the road to another set of gates. We sign in again. A sprinkler is at work, brightening flowers and greening a lawn. Through the gates, behind the works, we come to the treatment tanks. Rows of big, handsome Sandwich Terns sit in line along their edges, barely bothering to eye the car. They have black crests, like tufts of biros behind their ears. We carry on.

DANGER! SNAKES

BOOMSLANGS – COBRAS

PUFF ADDERS

'Crikey,' says Dad, mildly. I eye the edges of the track warily. How bizarre to realise that swallows, 'our swallows', spend half their time living happily alongside boomslangs and puff adders. If a boomslang bites, you bleed from everywhere until you expire. A good dose of puff adder venom will kill you in half an hour.

Now we are moving into the lagoons. Tracks run along dykes. Wild flowers flatten in the wind, the sea wind, coming straight off the Indian Ocean about 300 yards away.

Every time the path divides the choice is between a rough track and a rougher one, and DANGER! SNAKES boards stand sentry. There are flamingos in the ponds, and ducks, and then there they are, at last.

We barely recognise them: beautiful creatures with bright blue backs, the ferocious sea light whitened their undersides, they are like little arrowheads flinging themselves along ditches below the car, whipping up to our level, battling the ocean wind, peeling back over themselves, and even perching, just close enough to us, on a bush, for Dad to take a photo.

We jump out of the car, looking out for snakes, and meet the birds at the water's edge. They look different: smaller, paler and more ragged than the swallows I know. I had never seen one in moult. They begin to moult in South Africa, and many continue as they journey north, arriving in Europe completely refeathered, looking their best, ready to court and be courted. The extraordinary thing about the moult is the precision it requires. The migrating swallow must moult symmetrically: if a feather drops earlier or grows faster than its twin on the other side of the bird, flight will be unbalanced, manoeuvrability impaired and the chances of survival slashed. Every swallow is a collection of feather-bent parallel curves, growing in unison.

So though the swallow which landed on a little bush beside us was not a baby, could not have been, it did look like one. Perhaps it was a first-year female. Perhaps it was a first-year female born in our barn or above the front door on top of Mum's electricity meter, last May.

'Look at them – so close!'

'They are wonderful,' Dad says.

'Hello, swallow . . .'

'That's the one, is it?' Dad has a sidelong smile.

'Yes – Hello, sweetheart! Fancy a trip to Wales?'

We drove away, after a while, happy. It starts here, I kept thinking, here. And there is Table Mountain, and there is the sea, and that was my swallow. It has begun.

Two days later we went to the station, passed multiple checks to get to the platform, found my compartment and made our farewells. We made each other laugh. The train was very long and there were so few travellers that we seemed to be alone. He stood on the platform; I squeezed my head through the window. It is a strange thing to look at your father and know you are both thinking I might not see you again and this is just the way it is, this Friday the 1st of February.

'Bye Dad.'

'Go well.'

'I will. You too.'

We stared at each other for an instant, he gave me a nod and a look, turned, and walked away up the platform. Perhaps he could feel me watching his back: he went quietly, unhurriedly, and disappeared.

Right! I thought, with a jumpy queer feeling, this is it – let's have some fun! I opened and closed my notebook. Three pages were filled already, with Zeekoevlei. From now on I would write every day.

The train began to roll. I rattled around inside the compartment, changed position, stood, sat, pulled out my notebook and a pen, laid them aside and stared out of the window. Other trains went by as Cape Town began to give way to the flats. Just over there were Khayelitsha and Mitchell's Plain. At the edges of these suburbs the authorities have erected very high lamps, like prison lighting, and a row of tin toilets on waste ground, ready for more residents. People sat tiredly in the commuter trains which had doors missing; young men smoked and stared.

'High summer,' Dad had said, and it was. Too early for swallows to leave, of course – except certain males, often young and perhaps skittish, who cannot wait to set out, who drive themselves across continents, turning the migration into a giant contest – part survival competition, and part race: the first home will have the pick of the best nest sites, and therefore the greatest chance of attracting the best mates.

I watched South Africa all day. The train took a leisurely north-east curve out of Cape Town, following the route of the voortrekkers as they pushed north to escape the British. South Africa was first colonised by Bantu-speaking people from the Niger delta two and a half thousand years ago. They lived alongside native bushmen; Khoikhoi pastoralists and San hunter-gatherers. Portuguese sailors paused here on their way to the coast of Mozambique. In April 1652 a party from the Dutch East India company under the command of Jan

Van Riebeck founded a settlement where is now Cape Town, the purpose of which was to supply ships of the company on their runs to and from the spice islands of the East. Dutch, Germans and Scandinavians were joined by French Huguenots fleeing persecution in the France of Louis XIV.

The colony began to import slaves from Madagascar and Indonesia. Under apartheid their descendants, mixed with the descendants of their European masters – and the rapidly displaced Khoi-San population – would be known as 'Coloureds'.

In 1795 the British took the Cape, rather than risk its falling to Napoleon's France, briefly returned it to the Dutch in 1803, then took it over again in 1806. In 1820 five thousand British immigrants were shipped to the colony. The ruling White world of the Cape divided: the urban elite was now English-speaking, while the Dutch-speakers were largely farmers – 'Boers'.

The early nineteenth century saw the rise of the Zulu nation under Shaka, and increasing Boer dissatisfaction with British rule. In 1835 groups of Boers, accompanied by large numbers of servants, began to trek into the interior. These were the voortrekkers.

The story of the two peoples, English and Dutch, is written in the names of the stations; Cape Town, Bellville, Wellington, Worcester, Matjiesfontein, which is pronounced Makkeesfontane, after a spring, presumably 'discovered' by Matjie.

'If you get hot,' Dad had said, 'think about doing it with a team of oxen, a wife and family and all your worldly goods.'

In the east, in Natal, the voortrekkers fought battles against the Zulus, but victory brought only British annexation of the Natal region. The British in turn fought the Zulus, and began to import labour from India. Squeezed between native populations and the British, many Boers continued to press north. Eventually two Boer republics were formed: the Transvaal and the Orange Free State.

After the great trek this same route, which the train followed, became the road between the Cape and the diamond mines at Kimberley. Everything you thought you might need to make your fortune had to be lugged this way, and perhaps this was the way you walked back, after Cecil Rhodes had bought up your claim and defeat had emptied your pockets. Diamonds were discovered in 1869; Britain swiftly annexed the territory. In 1877 it also took over the Transvaal: a rebellion there led to the first Anglo-Boer War, which began in 1880 and ended swiftly with a Boer victory. The Transvaal became the South African Republic, or ZAR, led by President Paul Kruger. Six years later, gold was discovered in the Witwatersrand, a region of the ZAR. Now it flooded with prospectors and workers, black and white, and the population rocketed. In 1899 the British demanded that 60,000 non-Boer whites on the Witwatersrand be given voting rights. Kruger refused, countering that the British should withdraw their troops from his borders. The second Boer War began.

It was hot. The main street of Matjiesfontein is still wide enough for an ox cart with sixteen beasts in pairs to do a U-turn, Dad had informed me. I peered at it, dutifully. There were small, fast-flying hirundines near the station but I could not be sure of their species. The wind sighed gratefully in tall poplar trees. We wound up through passes and tunnels, along high, dry river beds, the train picking its way through the Drakenstein Mountains. Every now and then we passed massive stone blockhouses beside the track, built by the British during the second Boer War. Pretoria, the last Boer-controlled town, fell in 1900, but the Boers continued to fight for two more years. The blockhouses are enormous, with peculiar out-thrusts on the top corners, throwbacks to medieval castles, permitting defenders to fire down at attackers hard up against the foot of the walls. You would not fancy it. The Boers were notoriously good shots.

The British countered guerrilla warfare with a scorched-earth policy. To deny the Boers 'on Kommando' any support or provisions, their families, women and children were assembled in concentration camps. Disease, particularly cholera and tuberculosis, killed tens of thousands – estimates run to as many as 30,000. As I would find out, in many quarters this is neither forgotten nor forgiven.

The only other whites on our tourist-class Shosholoza Meyl train were thin, sun-bitten husbands and wider wives. They looked at an over-friendly tourist the way Londoners sometimes look at their visitors: with a blankness kneaded out of vague familiarity and vaguer irritation.

I slept in the long afternoon and woke towards evening, aware that we had been working our way up through tunnels. The pitch of the wheels had changed, there was a longer, running rhythm from the tracks. It was chilly, suddenly. I jumped up and gasped. I had read about it and seen pictures of it: the Great Karoo. The blue-pink colour of dusk, with red and blonde sand-streaks breaking it like waves, the Karoo was an undulation of short hard vegetation; under a bare sky its rolling contours looked as vast and cold as the sea. We had topped the escarpment and now ran freely on.

A flight of swallows mocked the speed of the train. Where would they sleep? Telephone wires? Surely they would not go down into the scrub, among the snakes. Were they on their way too? The sun purpled the Karoo and darkened its sandbanks; now the sea became a heather moor. A road appeared, made visible by the lights of a vehicle. For a while, whenever we kept pace with a truck or a car, the isolation seemed only to grow around us as our machines kept company.

The dining car was the first mixed-race place I had eaten in; behind me, a table of large men discussed business and put away pounds of thin chicken, fat chips and Coke. We stopped at Beaufort West in the early hours. Smokers drifted up and down the platform. A small boy did a handstand and the station lights were thronged with insects.

Kimberley was cool, briefly, at seven a.m. A man in tattered clothes who lived near the station and happened to be on the lookout for opportunities got out of his car (telling his dogs to shut up and his wife and a daughter that all was well) and drove me to the airport, where he was in charge of security.

'I do this because I have kids,' he said. He said a little about what it was like to be 'coloured' and talked about his brilliant daughter. Amazing to me, he seemed to pine for the apartheid era – for the certainty of its order, and the strength of its leadership.

'Things were better before,' he said, with an awkward laugh. So many said the same in the same way. The ultimate heresy is spoken all but freely, in a kind of chant.

'At least we knew where we stood.'

At the airport I sat in and out of the sun while it got hotter and waited while the hire car was made ready. The car, a rendezvous in Bloemfontein with a swallow expert and the visas were the only pieces of organisation I had put into the otherwise vastly empty map of Africa.

When it was ready I got into the car, a nondescript thing which made me feel like a travelling salesman, pointed it toward Bloemfontein, and drove. I set out across the vast emptiness of the Free State, feeling myself a tiny speck, smaller than a swallow in the gulfs of space which began at the roadside and lifted over flat land, which rose and filled with nothing but weather light and clouds as far as the furthest horizon. What a privilege and a pleasure it is to be alone in so much space! Tiny and inconsequential in the car, I had nothing to be concerned with, no obligations to fulfil. I felt a strange mixture of freedom and pointlessness. The self-containment of the solitary traveller gives you an other-worldly, off-to-one-side lightness of being. You have not the slightest bearing on events. You cannot even converse about the business of the day, supposing you have heard about it on the radio. You do not matter. The irrelevance of the traveller, your absence of responsibility, most of the time, for anything but yourself is a strange condition. You might as well be a ghost.

The country stretched away without break or end, its vastness echoed and dwarfed by the greatness of the skies. On the horizon low outcropped hills surfaced above the gold-green plains. Here and there was darker land: bare earth absorbs more solar heat, which causes thermals of rising air, but otherwise I supposed it would be as monotonous to fly over as to drive. To imagine a swallow flying across it all: to think ahead, to see the journey in its entirety would be to beat your wings against forever. They must look at the next contour, and surely anticipate the next change in the wind and the air currents, but do they see further? To the next night, the next roost, the next river? Or do they exist wholly in the present, propelled and pulled by urges of instinct?

Crossing the veldt the world seemed distant, as vague and

contingent as the day after tomorrow, but I came to Bloemfontein in the early afternoon. I became lost not long after reaching the town, meandering from black areas, where a market filled the road with an Africa I recognised – hundreds of people, colour, buses, exhaust, food stalls, cooking smells and the noise of voices – to white areas, where low houses hidden behind fences were laid out in a spacious, apparently deserted grid. I was rescued by a solicitous couple in a big white bakkie – as pick-up trucks are known here – who led me to a hotel named The Hobbit.

J. R. R. Tolkien was born in Bloemfontein, a fact of which parts of Bloemfontein are very proud. The creator of Bilbo Baggins and father of a billion units – books, games, films, toys – left town when an infant but still managed to create in his wake at least one hotel which encapsulates all the love of comfort, distrust of the world beyond the front door and the ineluctable yearning for adventure which you will find in the first ten pages of *The Hobbit, or There and Back Again.* My room was a bole of dark wood and white linen, with a green hue of sunlight filtered through clematis.

The television being jammed with provincial rugby – the Six Nations championship was about to start, England playing Wales later in the most suspenseful eighty minutes of the year, usually, to a Welsh fan like me – I hit the hotel's superlative collection of Wilbur Smith.

Wilbur Smith likes guns as much as the next man. He never misses an opportunity to reel off a full specification, including precise ammunition details. The impact of bullets on targets is recorded with a kind of moral faithfulness. He seems to see his role as a popular historian, with a mission to entertain. The struggles of black and white, English, Afrikaner and Zulu, from the First World War to the present form the backdrop to his blockbusters. He is anti-communist, sceptical of imperialists, and the heroes in his stories are always those who most love Africa. The continent itself is perhaps the true hero of his tales, embodied, generally, by some proud and fierce white South African with rippling muscles and a short temper. It is extremely rare to find a Wilbur Smith heroine whose buttocks do not remind their creator of ostrich eggs.

I woke to a call from reception, scrambled up and went down. Standing outside the hotel was Rick Nuttall, confined to the street by a locked gate which could only be opened from reception, Hobbit guests and their cars being spectacularly well defended. I recognised the large, handsome man by his friendly smile and his swallow T-shirt.

Rick is the director of the National Museum in Bloemfontein, and one of Africa's foremost authorities on Barn Swallows, though he would be too modest to make that claim: he talked reverently of Anders Moller in Denmark, perhaps the world's greatest expert.

We drove around town for a while, getting me my bearings and unravelling something of Rick's life and times. From a hill on the edge of town Rick showed me where north, south, east and west were, and we watched weather systems over the veldt. The light lay gold and blue and green on the grasslands, and, to the south-east, rain-black under a storm.

'I love bird-watching out there,' Rick said. 'Just going quietly, and stopping and staring.'

Rick speaks in that lovely, easy, irresistible-to-imitate South African accent beloved of British impressionists. You can hear his English heritage in it, his South African soul and the way he thinks. In my British ears African voices change English from water into wine.

The air was full of birds which Rick began to name: Pearl-Breasted Swallows, which shine like crescent moons (their dark, dark blue teaches you to see their fiery white); Greater Striped Swallows, which have longer tails and rufous speckled chests; Little Swifts, and many other flying creatures.

'It's like being born again,' I said. 'I know them all in Britain, pretty well, but here . . .'

I fell in love with birds at the age of seven or so: crows, ravens and buzzards were my first subjects. I began a list then, which all bird-watchers have somewhere, of all the species I could identify. By the age of fourteen the list was up in the hundreds, and I had seen my first rarities, like the Water Rail, a wonderfully retiring marsh bird, and Red Kites, which in those days could only be found in the very middle of Wales. Bird-watching abroad, on holidays to France or Italy, was always a disorientating business. Seeing most of the birds of Britain was one thing – something finite, which you could work towards – but on the continent you had to start again. And without knowing the culture of these new bird worlds, you could not really tell how excited you should be: Bee-Eaters and Rollers and Hoopoes seemed amazingly exotic and seductive, when I saw them in the Midi, but were they rare? Avocets were very special in Britain, but in the Camargue they were as common as magpies.

We went to Rick's university, pausing between students playing cricket and a reed bed to watch Red Bishops, fat little birds like shots of pure crimson. We sat on a terrace, moving under cover as the storm we had seen brushed the edge of Bloemfontein, and I filled the back of my notebook with Rick's expertise.

'We can track where they have been feeding by trace elements in their feathers, like iron, aluminium and calcium. The proportions present in a given area are found in the plants, and then in the insects, and then in the swallows. It's a wonderful thing because it is so unintrusive: you just need a tiny section of one pin' ('pin' from pinion, meaning feather).

The process is in its infancy because so little of the earth has been chemically analysed in this way, but there was a thrilled expansion in Rick's gestures as he described the possibility of actually mapping the journeys of a single swallow, and therefore those of thousands of swallows. Flipping open a laptop he displayed a hoard of data, an immaculately tabulated treasure, three years of results from his swallow-ringing project.

There were nationalities beyond nationality – Bloemfontein birds had been found everywhere from Cork to Transylvania, and then they had come back, and gone again, flying at speeds we can only estimate.

'Twenty-seven days!' Rick pointed at one line of a spread-sheet, 'Incredible isn't it? And that's assuming she was ringed the day she left and caught here the day she arrived.'

We marvelled at ages – first-year, third-year, sixth-year – distances, and death rates.

'Seventy per cent,' he said, soberly. 'In their first two years, 70 per cent mortality.'

I tried to imagine the corpses. Hundreds and hundreds of thousands of them.

'Oh yes, and it only takes three days of winter here to kill them.'

'Three days?'

'Yes, if they stay too long, by the second day of rain they have no energy to hunt, by the third they're in the mud and they can't get up. I get calls every year from farmers – they say "There are swallows everywhere, on the ground, have they been poisoned? What can I do?" The temperature only has to fall a few degrees, with rain, and that's it.'

We drove down to Rick's house, and he showed me around. There was a bit of wilderness, just beyond his garden fence, not too dangerous, not too developed, where he would normally expect to see swallows. No swallows. Here was where his wife and children would be, normally, though today they were having a brei – a barbeque – at his mother-in-law's.

'It's OK!' he reassured me, smiling quietly, as I worried I was dragging him away.

'My wife understands how important this is for me . . . talking about swallows, talking to other birdwatchers.'

We called in, briefly, on Rick's mother-in-law. I shook hands with his wife and said hello to their little girl.

'You're very lucky,' he said, where we stopped next, at a petrol station on a corner, opposite a house surrounded by trees, with swallows painted on the garden wall. 'There used to be so many of them roosting here that people would come in coaches to see them. Then they abandoned it, and we didn't know where they were, but just a couple of weeks ago I found them . . .'

'How many?'

'Impossible to say!' he laughed. 'About 1.8 million.'

'What!'

'Approximately . . .'

Later he said: 'I was taught by a great man. He said, if you want to be a good birdwatcher, when you hear a bird, go and find the bird. That way you will know its call.'

'And how did you get into them in the first place?'

'I was very young and I just remember looking across the garden at a dark green hedge and there was this beautiful thing, so amazing . . .'

He described a bird I have not seen, which sounded as though it was made of red and gold and green and white, the name of which I tried to remember, and on which he gently corrected me: a Double-Collared Sunbird.

We drove to a place near a river, with tall, tall blue gum trees. It was a brooding evening of towering clouds. The light faded as gently as falling leaves.

'Lesser Kestrels!' he said.

They came home in flocks, from very high up and far away, dropped down as though stooping on prey and settled themselves, sweetly, in barred ranks, like sheaves of little darts.

'I really must do this more often,' Rick murmured, as if reminding himself of something one cannot forget. I think he meant standing alone, with no thought for anything but the present. After a while we jumped quietly back into the bakkie and set off on our next adventure. Rick checked his watch, glanced at the sky, and frowned. Birds had begun to disappear and we had yet to see a single Barn Swallow.

'We've left it late,' Rick murmured, half to himself, as the bakkie accelerated. And so we were in a race now, as the light spiralled down and Rick spun us with increasing urgency through the suburbs. One moment too late and we would find nothing but a forest of dark reeds, screaming with chatter. Two moments too late and all would be silence. Three moments too late . . . Rick had another worry.

'It's used as a cut-through by all sorts of people – it's not exactly secure – so when we came ringing here a couple of weeks ago I made some calls and got some security. Just so that we could get on with it without worrying too much. We were very quiet but . . .' He did not look ashamed, just saddened by the truth of this. We were heading down towards a wasteland, the kind you find in any town anywhere, except here anyone could theoretically kill anyone and the world would have to say you had asked for it.

'. . . they may have . . . they may have gone down to the other side of the . . .'

Not the security: the swallows. The suburbs were thinning now and I knew we must be close.

'You just see them,' Rick said. 'There are a few people, students and colleagues, who would call me at dusk to say "I saw them going this way at this time", and then I got this sort of hunch and I looked at the map and I saw this place . . . I thought "I wonder . . ."'

We both kept bobbing our heads down to look up. I could not tell if he really thought we were too late.

'There!' he said.

'Oh yes! And there!'

As my excitement and disbelief mounted, his began to relax. The conjuror had done it.

'Another one – and there!'

'You'll start seeing quite a few,' he said, as I bounced in the passenger seat like a dog in a forest of flying squirrels.

They were coming in from all points of the compass now. It was not possible, it will never be, to know how many 'nationalities' there were. How quickly the air filled.

What is it like to stand under all those nations, all those experiences, under all those guesses, those eyes? You begin to try to see them all but you cannot see, you can only feel. Then, since guessing is impossible, you begin to know.

You know they have not finished eating. You feel the air devoid of midges. You hear the snap of their bills as they slide sideways, just missing your head, you feel the wind as one goes over your shoulder – snap! A bill shuts like a snicking trap-door. You half-hear, half-feel the hiss of the hunter's wake.

'When we were ringing we barely got bitten,' Rick grinned.

We had pulled up and stopped and walked a little way, over churned ground and debris. Ahead were the reed beds; water and vegetation covering an area about the size of two twisted rugby pitches. Behind us were trees, and, scattered round about, the fenced edges of different settlements. The air all around us stormed with silent wings. Sometimes we raised our binoculars. A man trudged by, not seeming to take much notice of the swallows, glancing blankly at us. Perhaps we looked like security.

And though no one killed anyone, all around us there was a mighty harvest of death. No swallow hit any other, of course, though they did not fly like starlings, or geese, or jackdaws, or waders, or any other species you can watch wheeling and whirring in thousands, in tune. Instead they seemed to delight in chaos, charging zig-zag into space which was at once empty and full, as though playing chicken with physics. They filled all the air our eyes could afford them in every conceivable direction. Our words deserted us again.

Chaucer, the poet at the fountainhead of English literature, puts the swallow in a strange place in relation to men and birds. His 'Parlement of Foules' is a poetic tour around the Garden of Love, with Chaucer as a visitor and the Roman general Scipio Africanus as his guide, at least as far as the gate. It is St Valentine's Day. Scipio pushes the worried poet (protesting he knows nothing of love) into the garden, where he finds all the birds of the world, summoned by Nature to choose their mates. Chaucer's tour of the characters and appetites of different birds is at once caustic, incisive and affectionate. In a broadly feudal, pyramidal food-chain (noble eagles at the top, countless seed-eaters down below) there is a hot, gluttonous cormorant, a 'waker' (watchful, wakeful) goose and a 'cukkow ever unkynde'. But among all the 'foules of ravyne' (ravenous raptors) and the lesser thieves, foes and destroyers of their preys, there is only one murderer.

> The swalown, mordrer of the flyes smale
> That maken hony of floures fresshe of hewe.

Chaucer places it between the nightingale, which he credits with calling forth the new green growth of leaves, and the 'wedded' turtle

dove, 'with heart so trewe'. Between the bird we prize most for its art
and one we idealise for its faith, in a couplet which encapsulates a
cycle of life and death, is the swallow, living by the murder of littler
things which here are positively angelic, doing no harm more than
making honey from fresh flowers. It seems a dreadfully human
predicament.

Now the birds were black flecks against dark twilight, white sparks
against black-green reeds, dull red blood-spots shooting close by us,
as if they were eating the light.

'Far side!' Rick said. He kept helping me direct my binoculars. It
was dizzying looking over there, across the dulled water to the reeds
opposite. Biblical plagues of the birds, denser than locusts, thicker
than blood-spatter, making a sound we could barely hear but clearly
see: a hissing, darting, scything thing, a terror. And then they began
to come down. Entire dark whirlwinds, funnelling down into reeds.
Was it fear, or thrill, or blind hive-mindedness that made them
unscrew themselves from the sky like that, so hectically?

'There was an Eagle Owl that used to come,' Rick had said, at the
abandoned roost. 'It used to just charge in and out of the flocks until
it wasn't hungry.'

'They're going, over there!' he exclaimed, and then mused, eyes
stuck to his binoculars. 'Perhaps because last time . . .'

The ringing again. But even so close to us, where the nets had been,
the swallows returned. We saw it and felt it. The reeds must have been
3 metres high and they shook and rattled, amplifying the rustling of
wings. At a guess – there is a birdwatcher's method of estimating dots
(you close your eyes and see how many you can recall in a glimpsed
area, then multiply the area until it covers the space you imagine the
flock filled) – eight hundred or so came down just to our left.

The reeds snaked away, thick as shadow, into the gloom that
surrounded the water. They chattered and swayed and gradually
silenced. A little wind and some insects returned. I got a bite and
stumblingly lit a fag. Rick smiled, blinking at me from behind his

glasses. He hauled out a cool-box and swept the tops off two beers. I toasted him in acknowledgement of this master-stroke. We had no fear any more.

Rick dropped me off at the Hobbit hole and we shook hands. I had no way of thanking him adequately but we had concocted a cunning plan to save face: roughly £70 pounds was worth 1,000 rand, which would buy a great many bird rings.

I had not dared mention, indeed I had forgotten, that while we were out England were playing Wales at Twickenham. I remembered now, and threw myself upstairs.

We had had a day, the swallows now had a night (and for some of them, the future possibility of that little manacle, human interest) and Wales, it turned out, had had England, superbly, freakishly, in the second half. The Supersports channels were so delighted with the defeat, rather than the victory, that they devoted a lot of the next twelve hours to it.

The timbre of their commentary was 'How the heck did the English have the nerve to face us in the final of the World Cup three months ago when they can't even beat a bunch of whippersnappers like Wales? Disgusting!'

My phone chirruped and buzzed with the celebrations of family and various friends. One called me, knowing perfectly well I was in Africa, and conversed as if I too was in a London pub and had just seen the game.

'Talk to you next weekend,' he said, happily, ringing off. We would be playing Scotland then.

CHAPTER 2

Namibian Roads

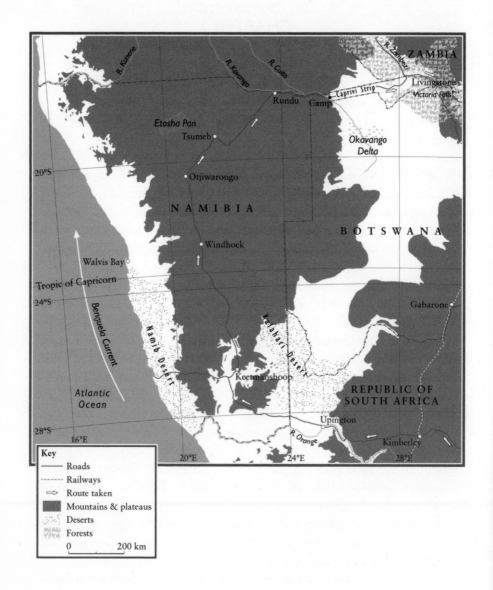

Namibian Roads

The Coucal is a portly bird with a long tail, like a fat cuckoo, and in Namibia the song of the Coucal heralds rain, they say. Two months ago, writing a piece for a travel magazine, I had been in a small plane flying over Namibia when the November rains came. The swallows would have been arriving then. We heard the Coucal in the morning, its bubbling, burbling cry, and in the afternoon we flew towards the storm. In the distance they were white clouds, blackening as we droned nearer. The storm-scope on the instrument panel flashed orange, then red. We pulled our straps tight, and then we were in it. Rain, rain like an elephant's legs, with a rainbow for a trunk. The land flashed, misted and darkened where the huge feet fell. We threaded between columns of water and thunder; the squalls were headless silver giants, 11,000 feet tall, striding westwards with walking-sticks of lightning. We flew down a dark tunnel through an arch of lightning.

In Zulu myth the underworld, hidden in the ground beneath us, and in the mountains, is the realm of cannibals. One story has a swallow carrying a bolt of lightning which breaks open rocks and frees the cannibals' captives. When you have seen the southern rain, the apocalyptic, smoking towers of water and thunder, it makes perfect sense that swallows, arriving at the same time, would be heralded as lightning birds; breakers of rocks.

Now I drove back across the Free State, then across the northern Cape, singing, through furious rainstorms, towards Upington on the

Orange River. When it rains in Africa the sea seems to fall from the
sky. Under the downpour at ground level you feel like a mosquito
under a pressure hose. The water crashes down and bounces back off
the ground in a kind of boiling mist. It seems impossible that anything
as small and fragile as a swallow could survive it. They must either
navigate around the storms, as we did in the small plane, or they must
take shelter. When the rains pass the insects come out and the birds
too, hunting and crying out, in a kind of survivors' banquet of
celebration.

I was happy alone now; I felt changed. I learned to count in
hundreds of kilometres in place of hours; I nearly died when a sudden
wind, a dipping corner and the speed of the car had combined for an
instant to spin my life like a coin. In Upington I found heat, real heat,
for the first time. I had shaved that morning; sweat and sun-cream
acidly scoured my face and neck as I drove in dazed circles in the
middle of town, road drunk and heat struck, cursing and unable to
make a decision.

'This is nothing,' said a local. 'This is not hot!'

I watched thousands of swallows going down the orange-brown
Orange River at dusk in a loose, vast stream. I tracked them as far as
the binoculars could follow them. There were massive mid-river
deltas of reeds down there; they could take their pick. I drove into the
Kalahari as fast as possible with the windows down, just to see it and
escape uncooked. Giving lifts to women and children I learned about
seasonal farm work along the Orange: all was well now, harvest time,
but in the winter their men just sat at home.

Going north from Upington, where I gave the hire car back, you
take the trans-Kalahari highway, which leads across half of Namibia,
to Windhoek. A long, long night on an Intercape bus began perfectly,
cruising at swallow height over the desert, past Snake Eagles on posts,
towards a sunset behind a thunderstorm. At the height swallows fly,
the world is a very different series of propositions, distinct from the
way we understand it, composed and threaded in ways invisible to
man. Villages are complete entities, towns are collections of districts,
of kinds of roofs, a graph in which few structures intrude into the

birds' realm. The way the world joins up, the way the land undulates through its features and under our impositions, are all legible to a swallow. Everything that is not of the air moves at a fraction of their speed, except vehicles, but even these are more limited and predictable than we conceive them, stuck to the roads. From the height of a three-storey building weather becomes much more visible and comprehensible than it is from the land. Every foot they gain in height will add miles to their horizon line, and therefore to the range of their forecast. They would have seen the thunderstorm we were now approaching long before I had.

It was a vision of coloured clouds, lightning and solar fire seen through smoking rain. Orange, pink and scarlet seemed to swirl behind the sweeping blue-black claws of falling water. The sky in some parts was a bare, stripped lemon colour, and pale blue. Lightning flashed and the lemon air seemed to turn green. We drove on towards it and everything darkened. The bus was going much too fast for its balding tyres, I had chosen the worst place, 'the suicide seat' (front seat, top deck) and my nearest neighbour borrowed half my minimal water supply.

We stopped for half an hour at the South African border while a young Angolan talked himself through. Then we broke down for another hour, in no man's land before Namibia, and sat in darkness, the bus rocking in the wind, lit sometimes by lightning, sometimes by the headlights of huge trucks rushing at us. The passengers sat still and almost silent in the darkness and the wind made the soo-op-wa sighing that gives the south-westerlies their name in Namibia. In the brief, brilliant light of the lightning I could see my neighbour's face running with sweat, as was mine. The trucks would pick us up, glinting in their headlights, light us up with the full beam and charge towards us: when they passed it felt like they missed us by inches.

It took our poor Angolan even longer to get us into Namibia, where we broke down again. Two passengers fitted new fan belts. The relief driver pushed the bus even faster; he was young and trying to make up time. In the end, dazed and furious, I took him aside in the small hours, at a service station at Keetmanshoop and said, forcedly smiling,

that I would kill him if he did not slow down, and that if he killed me
in a crash my brother would find him and kill him.

'Better late than dead, right?'

He was sardonic about it, and I did not care if it was only in my
imagination that we seemed to go fractionally slower after that.

The next morning a Norwegian called Dan and I drank whisky for
breakfast to celebrate our arrival. He was in Africa to teach fencing.
We passed the bottle and grinned tipsily at the Windhoek car park in
which the bus had dropped us. Dan showed me fencing moves as I
made calls on behalf of three puzzled young South African Catholics
who had come to train for the priesthood and had hoped someone
would meet them.

Lazing around in Windhoek, bemused by its heart (a German
shopping mall thronged with Africans), forswearing further buses,
organising a car, copying down official graffiti – One Nation! One
People! One Namibia! – listening to the crackle of gunfire from an
indoor shooting range, scribbling madly in my notebook and talking
to everyone I met, I began to feel that if South Africa was desperate,
Namibia was exultant.

'I love it here,' said a pilot, 'I'm staying.'

'Yes, it's good here. Safe,' said a taxi driver.

'My family have been here for generations,' said a hotelier. 'We
love it.'

'Oh, Namibia is wonderful!' enthused two overlanders. 'So easy!' A
doctor and an engineer, they had sold their house in London seven
months before and were on their way to Asia in an old Camel Trophy
Land Rover.

These were the voices of black and white, neither rich nor poor, of
German, South African, British, Namibian and New Zealand descent.
Any one of them, reading a cross-section of Windhoek newspapers,
would conclude that this was a country with as many doubts, problems, potential and actual shortages, as many troubles and menaces to
its future as any country anywhere. Even the most cursory survey of its
history – tribal archipelago, German South West Africa, League of
Nations mandate 'protected' by South Africa, revolution, war,

independence – tells a story as sickening and vicious as any on the great continent. The Herero and Nama peoples revolted against their German colonisers in 1904. Forty thousand had been slaughtered by 1906; an entire nation reduced to 20,000 refugees. And yet the contrast with South Africa was extraordinary. As though here, in spite or because of a gory twenty-three-year mêlée involving America, Russia, Cuba, Portugal, China, Britain, Zambia, Zimbabwe, Botswana, South Africa and Angola, the southern hemisphere had eventually seen off the northern, and black liberation beaten white spheres of interest, and anti-imperialism not merely survived but apparently overtaken imperialism, and something like harmony reigned.

SOMETHING LIKE HARMONY: A PORTRAIT IN OUTDATED STATISTICS[1]
Life expectancy: 52 years
HIV prevalence rate: 19.6%
Adult literacy rate: 85%
Estimated number of orphans: 67,000[2]

The size and emptiness of Namibia seem to question the very notion of what a country is. Namibia contains some of the oldest rocks on the planet; on my last visit I had climbed an upthrust of meta-morphic gneiss over a billion years old. Its brown skin cracked and tinkled like glass. Elsewhere in the Kaokoveld wilderness, the guide said, there were rocks that had weathered eight billion summers. Yet the country, the entity 'Namibia', has only existed since 1990. It is one of the least densely populated countries on earth, second only to Mongolia, with fewer than three inhabitants per square kilometre. On the Skeleton Coast I saw the bones of whales which had been killed by the whaling ships of the nineteenth century, scattered around the wrecks of ships and aircraft from the Second World War. It was as though the strip of sand between the murderous green surf and the bare desert was a graveyard of time. Our Land Rovers seemed as anachronistic as spacecraft, the span of our lives as brief and fragile as

[1]UNICEF to 2005.
[2]UN Human Development Report, 2002.

the mayfly's. Countries are political concoctions, and in political terms Namibia exists most clearly in opposition: it is what it is not. It is not South Africa. It is not a compilation of unrealised dreams and unfulfilled destinies, as South Africa is. Namibia is a *tabula rasa*. It is a vast ancient wilderness, in which people live, the descendants of a constellation of tribes: the San bushmen, hunter-gatherers, were the region's earliest inhabitants, followed by the Ovambo and Kavango, originally from the north; the Nama and the Damara in the south, and the Herero, who arrived from the north-west in the seventeenth century. In the nineteenth century Afrikaner farmers came up from the south, and from the 1880s Germans and other Europeans arrived, their descendants now forming, with the Afrikaners, a white strand in the Namibian nation. I met one of Namibia's older inhabitants, on my last trip. Her name was Kozombango, meaning 'of the caves', and she was a Himba, a nomadic race who came to northern Namibia three hundred years ago, fleeing tribal wars. She lived as the Himba have always lived, in a hut in the desert like an upturned beehive, kept company by a dog and a few chickens, her skin coated in a mixture of ochre and cow fat to fend off the sun. Outside her hut was an arrangement of stones which marked the site of the Holy Fire through which Kozombango kept in touch with her ancestors. The notion of Namibia, to this most Namibian of ladies, would be a laughable irrelevance.

In four days the nearest I had come to a swallow was talking to my landlady, who said she had one tattooed on her.

The swallow tattoo is a sailor's sign, meaning the wearer has travelled 5,000 miles at sea and returned safely. Some mariners are said to have had a second swallow done after 10,000 miles. The record for swallow longevity is eleven years: not counting the miles flown on the wintering grounds in South Africa plus all those accumulated during the breeding and feeding months in the north, that record-holding bird must have covered something like 96,000 miles before it fell to earth.

'We love them,' my landlady said. 'They bring the summer and the rain.'

Windhoek rain brought down seed pods like black boomerangs which crashed onto the roof of my room. Catkin and Gavin, the over-landers, retreated to the dining room and worked on their laptops. They were sleeping in a tent on the roof of their vehicle, which was parked in the hostel's forecourt. I wondered if I would look like them, one day: their bodies pared down to tones of goldish brown.

I entertained myself with a brief fantasy about buying a used car from the lot over the road – it had gull-wing doors and the owner said he could fix all the paperwork. It was definitely time to go. Road-testing my kit for the first time, walking a couple of kilometres through warm rain to find a hire car, I found it all worked perfectly. It was just much too heavy.

A BRIEF BUT SERIOUS LOVE AFFAIR

A little pale white thing, with a sense of style which reminded me of my adolescence in Britain in the 1980s, she was unprepossessing at first, all the more so when her carer said she was going to spy on me.

'If you crash or get caught when you're going too fast we'll know everything,' she said.

I was less than happy at her wailing when she thought I was in the wrong gear. But she was untroubled by the rain, did not distract me from finding my way out of Windhoek and kept her counsel on the motorway. Soon we were inseparable. There was nothing extraneous about her, except for her bleeped fussing over gears and revs, which afforded regular opportunities for conversation. (It is a treat to be able to shout something, every 50 kilometres or so.) She was so light and manoeuvrable that we were soon able to dodge flying bugs. (Harvesting blood and the bright splashes of butterfly wings becomes demoralising.) She reminded me of the huge family of sparrow-sized birds with crazily long tails, which cheep and twitter and seem to be everywhere, and though I sympathise with people who disdain people

who name their cars, I like company in solitude and preferred the
birds' name to the car's technical specification: she became the
Mousebird though she was a VW Golf.

On our three-day run north up the Trans-Caprivi Highway from
Windhoek to the north-east corner of the country, I learned the rule
of the Namibian road: read all signs, and take them seriously. In the
flooded centre of Okahandja the first HAVE YOU GOT A MOSQUITO NET?
appeared. I broke out the Malarone. In sudden, obliterating rain, I
practised holding my nerve and the car's line. We celebrated the
appearance of corners and their opportunities for steering as we
learned there might not be another for an hour. In hundreds of
deserted, green-yellow miles we became great fans of the lay-bys, and
the momentary friendships of the road; waves and smiles and flashes
of headlights.

　　Around us the bush flooded. The road was elevated a few feet above
the water, floating the bakkies, trucks and overloaded saloon cars as if
by magic over the impassable plain. Namibia's roads were made great
by the South African police and military. For 1,000 kilometres we
drove through battlefields, dating from before Nama and Herero
conflicts to the Cold War, over a heaven of fresh standing water that
was raining death and havoc across the continent. Namibia, Botswana
and Zambia were all suffering, the radio news said. Appeals had been
launched, rescue efforts were underway, but what are the efforts of
people, compared to the brutal power of the rains, and the great spaces
of Africa? All our powers are rendered as tiny as the kickings of an
insect. Nowhere I have ever been in Europe has the power to put us
in our true perspective in the dazing, almost chilling way Africa does.
Only the sea, perhaps, has the awesome indifference of the great
plains.

Hitch-hikers are the best thing about hire cars, I decided, having
conquered a fear engendered by the counsel of my father and his

friends that one should not pick them up. My first was a nineteen-year-old, his cheap jeans and white shirt immaculately smart: he was as neat and clean as a boy can only be at the beginning of term. He was a student of IT. My mission, following swallows, struck him as a bemusing eccentricity. It struck me as rather peculiar, too, as I explained it. The conversation led to Angola, where he had been many times.

'It's that easy?'

'Yes, I can go there, no problem.'

'I so wanted to go! But we need a fax from the Angolan Ministry of Tourism or something and I don't speak Portuguese and I think the guy I was talking to was stoned . . .'

'There are many British there – working.'

'Oil companies?'

'Yes.'

'So what's it like? I wish I could go . . .'

'The people are very poor. They have nothing but diamonds.'

'What?'

'If you go to Angola with a satellite dish, someone will give you a diamond for it. If you go with a car battery someone will give you a diamond for it.'

'Wow! Have you done that?'

'No! If they catch you on the border with a diamond you will have . . . problems.'

As we began to draw near to Tsumeb I thought I might like to live there. I had refined the tactics for melting space into time by changing my approach to road signs: I now studied everything except the distance markers, which I treated with great prejudice, concentrating only on their first figure. (Rundu 507 km is better seen as Rundu 5, as long as you then refuse to look at similar signs until they read Rundu 4, 3, 2 . . .) The problem with this is that the last 100 kilometres are less painful than the last 80, and so on, until the last 20, then the last 10, which are an agony. But coming into Tsumeb my spirits lifted with the changing geography: hills as round and smooth as the secret heart of Wales sprouted succulent trees from fair grey rock.

Tsumeb is world-famous among mineral collectors. The head frame and winding gear of the lead mine commands the centre. Behind Main, a grid of lesser streets were streaks of red mud and black tarmac and the trees seemed full of flowers. I went for a wander, fled from the rain, burst into the sports bar, spotted the Wales–Scotland match in progress, exulted at the score and approached the bar.

'May I have a beer please?' I asked the barman, who was so self-effacing you could hardly believe he had heard and so quick to serve that you felt embarrassed for having asked.

'Where the fuck are you from?' demanded a white man beside me.

'Wales!' I cried euphorically. 'And we're winning!'

The fat, pissed farmer looked unimpressed and I did not give a damn. I would have him and all his mates, if they wanted it – or at least I would outrun them. He sniffed and shrugged. He could have squished me on the floor with a finger.

In the Tsumeb sports bar they have a stuffed baboon, a hideous thing, wearing an England rugby shirt. The television shows rugby but nobody watches it when obscure European teams are playing, unless, one imagines, England are being thrashed by a superior southern hemisphere side.

I loved Tsumeb, and I loved Namibia, and I felt only benevolent curiosity towards the other specimens clinging to the bar, who would possibly have punched me for my English accent. To hell with them and whatever they thought, we were only twenty minutes in and Wales were definitely, probably anyway, going to win again! (Thank God for Scotland.)

And hurrah for Namibia, I thought, hurrah for the black revolution. This is exactly where these white bastards should be, clinging to a bar, murdering each other's livers and lungs, with a black barman who is too clever to ever rile them; let them rot in here, let them die hard and bitter where they can do no more harm. Let the future belong to the black couple I saw outside, young and beautiful and clinging to each other, shrieking and running from the rain which came sizzling across Tsumeb's lovely public gardens like a giant green-grey carving knife, and drove me in here in the first place. I had been on the road for a week and a day and seen 1,000 kilometres of Africa, and today was not for defeat.

Taking in the way they treated the barman, and the way they obviously felt about me, it took less than a minute for my liberal principles to dissolve in a declaration of mental race-class war on an entire section of southern Africa's natives. But then one, a wiry man with a buzz-cut head, began to talk to me. He was a few pints in when we started and I was soon downing bottles on an empty stomach: we became rapidly matched in drunkenness, and though red shirts battled blue shirts behind him on the screen at the other end of the room, we joined gazes and intimacies like old pros. He really was one.

'Of course I fought in the war, that's the difference between you and us, we've done it, and – I know this sounds – to you – whatever – but we understand them, you know? I *know* the kaffirs' (he failed to whisper it quietly and the barman did not even flicker) 'and we can work together, yeah! Fine! Because I know them, and they took my fucking farm – well, bought it, yes, but they didn't give me what it's worth, not a shred – what? Now? Farmer. We farm veggies now. Fucking rain, I'm telling you! Aren't we, babe? Farmers! Hey! That's my wife. Hey! This guy's all right, he's from – where are you from? What? What was it like? Nah, give a shit, I'll talk about it, they didn't give us shit. They let us get beat. They let us fucking die there. Fucking South Africa, my fucking country, I'll never go back there, it was a joke – we weren't getting petrol, we weren't getting planes, they let us die so they could pull out . . . What's that? Oh yeah, Wales, eh? Hey, are you coming back? You're coming back? Hey, Bronckie, you should meet this guy, he says he's from Wales. After the second half? Later? Where are you staying? If you need somewhere to stay you can stay with us, can't he, babe? Babe! Yeah, right, see you later, take . . .'

I cannot even recall the second half of the match but I remember something he said: 'We were fighting for you.'

As for Bronckie, he mostly mumbled, but the gist was clear again.

'History' for him meant the Boers discovering South Africa, the British putting their wives and children in concentration camps (300,000 died) and stealing the country, and everything since flowing from there. The wars I was raised with did not seem to have happened; the First and the Second disappeared, the Cold was neither here nor there. He too was kind to me, in his restrained distaste, considering that I was a Nazi child to him.

Leaving Tsumeb the road to Grootfontein curled around to the south before heading north-east again, winding between pasture and hills. The pasture was high-fenced and thick with thorn. Now and then a gateway showed tracks or roads to ranches set back from the road. The world and its children were out, farming: five sweating herdsmen ran uphill after fleeing cattle.

I stopped to take in the road and the space, the birdsong and the feel of the morning. A wide tarmac strip with a broken yellow line arrowed away from my feet to the horizon. Swallows were using telephone wires as a base, hurling themselves into the rain-freshened air and zapping about in the sun. I had seen them now and then on the drive north, but only fleetingly. This was the best view of them I had had since South Africa, and it seemed a validation. There were round hills and high clouds in schools, and a great brightness in the day. The collective noun for swallows is 'a flight', which seems a little pedestrian, compared with a murder of crows, a convocation of eagles and an unkindness of ravens. An exaltation would have suited this group, but larks have bagged that. A confirmation of swallows, I resolved, would be my private term for them.

The passing of rain bought dung beetles onto the road, and a little later, gliding low over the car, yellow-billed kites. Hundreds of yellow-billed kites! You jink for dung beetles – they are motoring too, generally at an angle to your line of travel – but you slow right down for kites. Kites are magnificent soarers, gliders and turners but when they are panicked they flap. Compared to a swallow's mastery of speed and superlative judgement of changing depths of field, kites are flying babies. To have one come through the window at even 30 kmh would be mutually fatal. It was their moment and their road and it only lasted for half a kilometre, and they were spectacular, but I do not believe I was the only driver who swore at them.

At the first service station at Grootfontein all the newspapers were leading on the floods. Poisonous-looking expanses of brown, outbreaks of malaria, communities cut off, aid unable to follow the photographers. Travellers, loafers and all their attendants milled around the forecourt.

'The problem with Grootfontein is nobody stops,' said a boy. 'You are all going through.' He was about sixteen, with a quick, attentive look and a slower, philosophical smile. He wore the same cheap clothes that all those hanging about the forecourt wore. I wondered if he would simply stay there, hanging around Grootfontein, until he became one of the hunched old men sitting on the low wall by the road.

'Yes,' I said. 'Would you like anything from the shop?'

'Oh! I will have a Fanta. Thank you.'

I went in for a paper and drinks.

'What do you do?'

'I am trying to save money to continue my education.'

'How?'

'With my art.'

I looked down. I had thought a Fanta was a good way of buying my way out of buying something I thought I did not want. My mother and brother had visited Namibia a few years before and brought me back presents: a bracelet, made of wood and little discs of ostrich shell, and a carved Malakani nut, sometimes known as vegetable ivory. You find them all over Namibia, often decorated with animals, as this one was, and topped with a looped leather bootlace. He was doing it as he spoke to me, curling black marks onto the white and brown sphere of the nut. As I watched, in the time it took to swig a drink, a zebra was made out of stripes and a giraffe finished off with spots. The animals had an angular, electric quality. A critic might have seen echoes of cubism. Leg-scrapes had muscle and kick and line. I had decided a while ago that I would not, could not, fill my rucksack with trophies. Art was a different matter.

'Rundu', said the guidebook. 'Steamy . . . tropical . . . across the river from Angola . . . cashpoint . . .'

It would take the other half of the day to get there. In the middle of the afternoon an arch loomed over the road. There was fencing on either side of it and uniformed men milling around. I crawled up to it, trying not to look suspicious.

– – – – – – – – – LINE OF DISEASE CONTROL – – – – – – – – –

The men manning it were very efficient and cheery. One took a good look at the Mousebird and I. We exchanged greetings and he waved us through. We rolled over mats which may or may not have been soaked with chemicals and drove into a world extraordinarily changed.

The ranches had gone. The fences had gone. White and black Africa had gone, rich and poor Africa had disappeared from view. If there had been well-off families, or farmers with thousands of acres of beauty, wealth and worry, employing dozens of labourers, or great homesteads somewhere out of sight beyond the road, then there was no sign or trace of them now. The world as I understood it had gone. There were no more agricultural bakkies. The silence of the sky and the songs of the bush were gone. Now there were men, women, children, huts, goats, sheep and cattle. There were many schools and many more shebeens. There were convenience stores, one-room garages, little kraals of thorn, rough circles of homes, straggles of bush, tall trees, where bee-eaters and lilac-breasted rollers perched, and skittering children trying to keep goats off the road. To the south of the Line of Disease Control, cattle could be commercially farmed, sold and transported. Here, to the north, livestock could not be transported south, so something not much more profitable than subsistence farming had replaced agriculture as the west understands it. The Line of Disease Control seemed to have cut time. I saw no more whites.

I guided the Mousebird down the middle of the road, two thirds of her in the left lane and the rest in the right, hoping to maximise swerving options. The days of 120 km/h solitude are replaced with second-to-second choices about eye contact, waving, slowing and speed. There is also a great variety of hitchers. I picked on Danny because he was alone, a little way past a busy area, under a tree. When he saw that I was a single white man he stopped waving and smiled uncertainly.

'Where are you going?' I asked.

'I am going towards Rundu.'

'I'm going there. Jump in.'

'Yes – but I have a goat.'

'Bring your goat!'

Danny is not a very big boy. He takes the forelegs and I the hind; both sets are tied together. The goat is beautiful, chocolate-brown with cream and dark eye makeup. It cries out pitifully as we manoeuvre it into the boot. I am used to sheep in the backs of cars but my only dealing with goats was a white, long-haired young brute called Shenkin, mascot of the Royal Welch Fusiliers, which I was sent to interview as Work Experience for the *Western Mail* in Cardiff – Shenkin and the band of the Fusiliers being part of the pageant of Match Day. I questioned Shenkin via his handler and then took a breather, aware that lighting up on the hallowed turf of the old Arms Park was about as fat a payment as Work Experience could offer. When next I considered Shenkin he had my notebook in his mouth.

'This is going to be a good meal, isn't it?'

'No, this is a female. She is going to have kids.'

She is much better behaved or more fatalistic than a sheep. Once we have shut the boot, plunging her into darkness, the heartbreaking cries fall silent and there is no banging.

'Where are you going with her?'

'I am taking her home . . .'

'You just bought her?'

'Yes, today.'

'How much was she?'

'Four hundred dollars.'

(About £25.) 'Is that a good price?'

'Yes . . .'

I thought Danny was reticent because he was so young and shy, and because I was asking questions too curiously. He had a rare, slightly strained smile.

'My father has died. We are having the funeral.'

'Oh God, I'm – sorry. Did he die recently?'

'Yes.'

'Do you have brothers and sisters?'

'Yes but my brothers are away.'

'So you are the head of the family now?'

'Yes . . .'

He is not yet thirteen, I guess.

'How did your father die?'

'He had a pain in his stomach.'

'Did he go to hospital?'

'Yes, but he died.'

'I am very sorry, Danny.'

The boy ducked his head, smiled faintly and continued to look out of the window on his side. We passed another shebeen. They are small cubes with dark interiors in which beer is sold, not much bigger than the table football you can sometimes see through the door.

'Are you a tourist?' he asked me.

'Yes . . . sort of. I am a writer. I am following swallows.'

'Following what?'

'Swallows – little birds, they come in the rainy season? You know, little blue and white birds, they fly very fast, like this [zig, zag] . . . if I could see them I would show you them, if we keep an eye out we might see some . . . they spend your winter in Britain, my home, then they come here for your summer.'

'Yes.'

After a while I said, 'Your English is very good. Do you go to school?'

'Yes, thank you. I learned it.'

'At school?'

'Yes.'

'And do you still go?'

'To school?'

'Yes.'

'Not really.'

'Because you are very busy at home.'

'Yes, very busy . . . my home is here, by the big mopane tree.'

'Right! Stop up there then?'

'Yes please.'

We turned under the mopane and stopped. We lifted the goat out of the trunk. I was amazed, almost disappointed, to see that she had not soiled it; any ewe would certainly have left some tokens but all that remained of the goat were a few fine brown and white hairs. There were some huts set back from the road about 50 yards.

'Is that your house?'

'Yes, they are having the funeral there.'

'Do you want me to help you carry the goat?'

'No . . . someone will come.'

'OK then.'

We both hovered for a moment.

'Danny,' I said, suddenly, 'you're such a bright boy. Please – you must try to go to school . . .'

He looked at me with such a strange look, something like pity. We shook hands and smiled. I drove away.

I came to Rundu in the late afternoon, apprehensive about it and excited. It was the gateway to the Caprivi Strip, a region about which

the guidebooks had been tentative and gnomic. It was closed off by the
Border War (when South Africa fought the Marxist liberation movement in Angola), then opened to convoys, then shunned again when
four French tourists were killed in 1999 (supposedly by UNITA,
Jonas Savimbi's Western-backed, mercenary-rich, anti-Marxist,
above all pro-Savimbi militia, long after UNITA had lost their fight
for Angola.) The strip itself is one of the most wonderful colonial
perversions.

Even by the insane, bandit laws of imperial cartography, Namibia
ought to be a rectangle. The South Atlantic and the magical,
diabolical Skeleton Coast to the west; then the Orange River, more or
less, to the south; then a straight line up through the Kalahari – jutting
slightly to account for the Okavango swamp and Botswana – takes you
to the horizontal northern frontier with Angola, formed by the
Kunene River, home to the most ferocious crocodiles in all Africa, and
the Kovango River. But in 1890 Queen Victoria's men gave a long flat
pencil of land running between the Kovango and the swamp to
Bismarck, allowing him to link his vast, boiling wastes (the Kaokoveld
desert is a truly extraordinary and still near-impenetrable world) to
the Zambezi, which runs eventually to the sea on the other side of the
continent. It was not a gift but a swap: Victoria's subjects got
undisputed claim to Zanzibar, prince of spice islands, a prize so rich
her ministers also threw in Heligoland – some North Sea rocks – as a
sweetener. And Heligoland surely swung it: on paper it is otherwise
hard to see how it was a good deal for Germany, partly because of the
small matter of Victoria Falls, at the end of the strip, where the
Zambezi commits suicide in a giant inverted volcano of sparkling
tortured water and roaring rock, which effectively stymies navigation.

The upshot is that Namibia looks like a child's attempt to make a
paper rectangle, a distracted child, which did not bother to tear off a
last strip jutting off the top right corner: Caprivi.

Rundu is a left turn off the main road, low buildings, lots of people
and heavy heat. The guidebook says something about a place to stay
by the river. The main street has a curl, a division, and then you hit
the river road. There are various signs promising lodges. The road

down to the one I chose was rough: the Mousebird coped easily with
its gradients, broken surfaces, floodwashed gravel and holes, but
though we did it several times there was always something interesting
between the top and the bottom – a half-nasty spin, a scrape. I missed
the turn and ended up at the beach. There were two or three cars and
one or two people, listless in the heavy, yellow, riverine peace. I asked
someone, who pointed just uphill. We spun around, scuttered back up
and turned hard right through an open gate onto a sand track. There
were bushes, little lodge huts, a shuttered bar/breakfast room almost
overlooking the river and no sign of anyone at all.

There was a choice between full sunlight or slightly too cool, too
deep sand, so we stopped where we were. The owners' house ('*owner-
occupied, delightful German couple*' or something,' etc.) was set back in
bushes a little way above the gate. I walked slowly. The front door was
shut but everything else was open; a radio played along with the
breeze. At just this height above the water in just this much shade it
was halfway between very hot and perfect. This is exactly where the
colonisers always seemed to live – in the speck of space which nature
has made most appealing to people.

A girl came eventually and checked me in, unlocking the bar,
pulling out a book, filling in a line, giving me a key to a hut. The floor
cooled my feet; there was a double white bed, a huge mosquito net and
a television showing ghosts replaying football in a storm. (African Cup
of Nations – the final is tonight!) She was in her early twenties,
perhaps, and bemused. I feared she did everything. I was loathe to
unpack, to dirty the loo or derange anything, but I did, a little, starting
with the aerial cable to the TV. The ghosts disappeared but not the
storm. I went looking for food.

The supermarket was playing host to a small party of exotic
foreigners. Nobody else seemed to pay them much attention. I tried to
work out where they were from, which meant eavesdropping. They
were better-looking than they are supposed to be and younger than I
had been led to believe was now the norm. Their truck looked good.

It used to be said that the overlanders partied their way around, sleeping it off as Africa rolled and bumped tiresomely past. Now, it was said, the demographic had changed, and the trucks were full of retired Adventurous Travellers, beadily sharing knowledge about humans, humanities, earth sciences and is that or is it not a Bearded Vulture?

These looked businesslike, buying supplies, and sexy, in their cottons, straps and tans, wearing that quasi-pained, pointedly-not-pushy expression you might see in Sainsbury's on a Saturday. They were of a type, and I and many others in Rundu's supermarket stared surreptitiously at white women.

It was a strange meal; self-catering courtesy of the ladies behind the meat counter at the supermarket: a warmed chicken leg, some sort of bread and something to drink, taken on the veranda. Every time I dropped a crumb a line of ants switched direction and picked it up. I went for a walk by the river later: not very far, because it was hot, and cautiously, just because. Later still I went for another drive, vaguely wondering where the sewage farm might be. Angling the car away from the river, gracefully switching sides to avoid schoolchildren (many schools do two or more shifts, so there are three big change-overs, morning, noon and night) I came upon the millipede.

Blacker, shinier and meaner than an escaped bicycle tyre, it was trundling in a dead straight line three-quarters of the way across the road, on our side. Three impulses hit me at the same time: public service (take the monster out), self-preservation (offer it a drink?) and something else, which escaped me in a whistle of amazement and made the Mousebird veer wildly.

I am not sure if it was the millipede, the final of the African Cup of Nations, the threat implied by the vast and beautiful double mosquito net or the amazing near-silence, but I left that camp and drove through the gloom to another lodge, a bit further down.

Around the television the atmosphere was unhappy. Cameroon, the Indomitable Foulers, were playing Egypt, the Indefatigable Pounders, for the championship. Regardless of the fact that Cameroon wanted it so badly that all of West Africa, southern Africa, and indeed Black

Africa wanted it too, the barman, being Namibian, was disgusted with the whole thing, because South Africa had underperformed so thoroughly that they had contrived to be knocked out in an early round – which rather killed the atmosphere in the otherwise empty bar.

Mine was a hot wooden room, beyond the wall of which was another solitary man in an identical room. I ate warm rice, noted that some sort of local government conference was taking place and tried not to laugh, the next morning, as the black staff took their wonderfully lethargic yet painstaking time misimplementing the harried orders of their white manager. His anxiety should have been catching. Without anxious hard work, you could see him agonising, the conference attendees would not get the five-star service to which they were entitled, regardless of creed, colour, price or facilities. Yet his staff simply refused to contract his anxiety, or work at anything like his pace. And then there was rain.

The heat of Rundu, the proximity of the river and the thick green morning were completed: the rain was all you could really hear and all you wanted to look at. Even 3 feet into cover it caught you; huge drops bursting into smaller drops, billowing into spray. It crashed into us and splashed under us as we headed down the Caprivi Strip. The plan was to go just under halfway along it, turn right and cross the Botswana border, gateway to the famous Okavango Delta, a swamp so enormous that there simply had to be swallows there.

Instead, by following the Mousebird's nose down the strip, following signs to Botswana and taking a left, we became first distracted by, then entangled with, an extremely poor road.

Barely 40 yards into the bush the track disappeared into an untroubled pond of brown water. We battled on. The paths divided in an unsettling fashion, sandy high road versus vegetable low road, but the Mousebird would not daunt. Some very large vehicles had done that track but I would not have been in one, not with the narrow causeway before us now. Something which looked like a collapsed bridge passed, then we came to another, which two men were building. Over a hill, down a dale, and we were in. The track turned into a soft loop. There were thick trees, the wooden wall of something that promised a bar, and beyond that the unmistakeable presence of a river. A huge green truck was parked in the shade, beside a couple of sober green bakkies.

The Mousebird was now modelling a sable-yellow leopard-print with matching tyres. Her white undercoat showed it all up to perfection. I was barely half out of her when a white man appeared, raising an eyebrow. He was older than me, quieter, and, I was perturbed to see, was not driving some sort of Landcruiser, which would have instantly given me one up on him, but the slightly tatty kind of estate favoured by nice families with small children. Worst of all, he appeared to have just done the shopping.

'Hello,' I said.

'Hi. Where have you come from?'

'Cape Town!'

He smiled.

'Not in that, you haven't.'

'Ah, no – Windhoek . . .'

We shook hands and introduced ourselves. His name was Mark. His wife was Margie. This was their camp.

'You wouldn't have any beds, would you?'

They gave me a mattress in a tent standing on a platform in a low tree down a narrow path, by the river. It was a lovely spot, and it had a snake.

'What kind of snake?' Margie asked, when I remarked on it.

'Green! Quite small, about so long, thin . . .'

'Oh, it's probably a Western Green,' she said. 'Harmless. What shape was its head? Coffin or diamond?'

It did not act like a snake. They are supposed to withdraw when they hear you coming, I believed, as I stood on my platform, watching the beast ushering itself into a bush of dry twigs about a yard from the tent flaps. The twigs did not make a sound and the snake stopped, poorly concealed, now you knew where to look. It formed a long coil, with the tail disappearing into the bush and the head pointing back out of it.

'Knob off!' I hissed. It seemed a silly thing to say even as it came out, not my sort of curse at all, normally, but then I had never confronted a snake before. I made a gesture, some sort of noise, a stamp of the foot and another curse – and it came at me. Well, it moved, and not backwards.

'Er, coffin-diamond?'

'Well, it's probably a Western Green.'

'What shape do –'

'Diamond. But there are these things called green mambas around . . .'

'Green mambas.'

'Yes, and they have coffin-shaped heads.'

Perhaps it is inevitable that such a place should have a special dog. Her name was Slim, she was fat and no fan of crocodiles. Her sister, Shady, had been eaten by one. One of her rivals for human affection, a rangy bitch often the butt of the camp's paint-ball gun, had had six pups in the back office. Three of them went down a rock python.

In the Kovango River, they told me, lives the Dikongoro. The Dikongoro is a dragon. What does it look like, I asked?

'Well, it has horns like a dragon, scales like a dragon, legs and claws and teeth like a dragon, the tail, the face, the eyes of a dragon . . .'

'The wings?'

They were not so sure. 'More like a Chinese Dragon,' someone said.

The Kovango River at this point is wide and browny-green. It is warm and silky but you had to be careful about swimming in it, because it was high and strong and hugely deep; after the rains it rises and turns banks into islands, islands into reeds. All the huts near the Kovango here are on high ground or on stilts, and the difference between high ground and marsh depends entirely on the rain that falls in Angola and other places, and so varies from year to year. There are crocodiles and hippopotami in the Kovango, which are two more reasons for being careful and quiet, but the best reason of all is the Dikongoro.

Why, I wondered, does everyone speak so quietly of the Dikongoro?

'Has anyone seen it?'

They fell silent. This is the trouble with this dragon, and the rather wonderful thing about it. If most dragons see you, they might eat you. Not the Dikongoro – no one mentioned anything of the sort. But they did say something else.

'If you see the Dikongoro, one of your relatives will die.'

Really, I said, has anyone . . .

'Yes,' they said. (And now they whispered.) One or two people living around here did see it, and relatives of theirs did die.

Really, I said, who?

'There is one guy who works here, whose mother saw it . . . and another woman . . .'

And then I really did not want to know too much more. It was very clear that as a guest I could certainly ask around and meet local people who had seen the Dikongoro. Indeed, I could and did talk to several of the camp staff whose relatives had seen it, but we did not talk too much about such things. Instead we talked about the dragons themselves. Some said there was one, some said there were many. There is a certain island not far away where the Dikongoro live, and depending on who you talk to, the Dikongoro is either there or not there, and either minds if you go there or does not, and she is either a female, a mother, or she is many. Some even said the island was where the Dikongoro go to die.

No doubt the story of the Dikongoro is almost as old as human settlement on this part of the Kovango, and the locals among the camp's staff took it very seriously. The strange thing was the way their belief affected the white staff, all of whom mentioned the dragons, with a laugh, but then dropped their voices and said, 'But you know they really do believe it. To them it's absolutely true . . .' in such a way that you looked uneasily at the river, and a bit of you believed it absolutely too.

The core of the white staff, three South Africans and two Europeans, were around my age and included Byron, from Pretoria, who was rarely seen in anything but underpants. A birdwatcher, a naturalist, with the muscles of a man and the quick smile of a boy, resplendent in dark curls and blessed with no apparent traces of fear whatsoever, Byron was most happy fishing – in underpants – zipping off into the bush to do something – in underpants – and leading tours.

'Are you coming tomorrow then?' he asked.

'Sure – where?'

'For a bird walk?'

'Are there any swallows around?'

'Hmmm – maybe, yeah. Up by the village you sometimes see them.'

I sat by the Kovango and wrote. How does the white man come to Africa? As he always has. As a hunter. As a missionary, evangelising Development in place of God. As an adventurer, seeking near-misses to prove his existence. As an explorer, searching for miracles to take home. As a trader-plunderer, with hard currencies on his side, instead of the rifle's guarantee. As refugees from the world we have made; as boys, resolved to become men.

The camp emptied of other guests, and the staff and I had it to ourselves. There was a floating cage on the river, for sunning and cooling off and not being eaten; there was a shower to redecorate; a staff auction to organise; there were meetings held – a post-mortem of a recent canoe trip – and someone had to go to town to pick up some diesel and other treats.

Mark, whose brain-child all of this was, is keen that guests should not miss the souvenir shop. He is a South African, an engineer who had spent much of his professional life building roads in Botswana, he said, which seemed ironic, given the appalling state of the road to his camp.

'Ah,' he said. 'But that's no accident.'

The souvenir shop's stand-out item was not for sale: a long polemic on six sheets of paper, pinned to the wall, called 'Mister Westerner'.

I had a look, then he summarised it for me.

'It says, Mr Westerner, you are welcome here, but stick your hands in your pockets and keep them there, we don't want your money, and shut your mouth, and look around and learn.'

Mark was looking forward to the crash that the world's media was then predicting that world was about to face. As far as he was concerned, Caprivi and the environs of the camp had been miraculously conserved from the predations of modernity by years of war, and now represented an opportunity for a redrawing of the terms of business between locals and incomers.

'You've got kids here who go away and come back and now won't go near their parents because suddenly they say they smell. Who don't respect their parents because they live in a hut, not a house. Do you know what the Human Rights Act did? It killed the traditional structure of law-enforcement round here – they used to just take you into the middle of the village and hit you with a stick: not any more. So the old respect is going.'

'So what do you do?'

'You give everyone a basic standard of education, to the point where they can make an informed choice: do I want to join the rat race or do I want to stay here? But what you don't do is tell them that one is better than the other . . .'

He was planning to defend the area with three strategically placed checkpoints. 'There will be a hut, with display boards showing how people live here, and something like Mr Westener: we don't want your bloody aid here, your bloody values! We've already got the values of —— here!'

'So this is an environmental re-education camp masquerading as a holiday?'

He laughed. 'Look, I'll show you something.'

It was a tree stump.

'What's this?' he asked.

'. . .?'

'I say it was the tree of knowledge. And when I came here I cut it down.'

All the other trees and the grass needed a lot of river water pumped up to them, and soon he went off to see about it.

Byron appeared in his underpants with Cristoph, who was as tall and lean and black as Byron was brown: they were like brother warriors; they looked as though they could run all day, catch their supper and disappear into the vastness without a care.

'Bird walk?'

'Yes!'

'Might get a bit wet. OK?'

'Sure.'

'And it's still very hot. Have you got sun cream?' Margie asked. She would be coming with us. She was wearing her sun hat. We set out.

We left the camp and bent left, following a path through tall golden grasses. We passed huts and a stockade; a village grouped around a great tree. Soon the path was twisting gently through reeds. Byron and Cristoph veered off to the right.

'Stay that side if you like,' they said. Now our path ran through shallow water, while theirs plunged them in up to their thighs.

'Blacksmith Plover!' they cried, shooting out hands like spears.

'Right! Lovely!'

It was a golden evening. The water was beautifully cool, and in some places my feet plunged into black silky mud, cut with sharp reeds. Byron and Christoph were in quite deep now, splashing confidently.

Byron paused: 'Crocodiles?'

Cristoph laughed. They shook their heads and plunged on.

'When the rains come there will be hippos and crocs here,' Margie said, as various ducks, geese and waders got up, were named by the men, and disappeared. She told stories of expeditions with Mark. Their favourite place was Mana Pools, she said, a reserve on the Zambezi in Zimbabwe where she once looked up and saw a leopard gazing down at her from the branch of a tree.

'Mark and I have a deal, which is if one of us is killed by an animal, no one is to kill that animal.'

'Even if it's a croc?'

'Of course!'

There was a commotion in the reeds. Suddenly Byron and Christoph came haring past us, crying out.

'Great Snipe!' they shouted.

The bird looked like a fat woodcock. It burst from cover only moments before the hunters went hurdling after it. They marked its flight and charged at the spot where it went down. We were all in up to our thighs now, there was no path and no boundaries, just trees, outcrops of higher ground and the evening light playing over all like the wind. I was carrying my shoes and giggling at things.

'What is it?' Margie asked.

'It's those two – like mad hunting dogs! And I was just thinking I got a text message from my brother today he said hi hope you're having a great time Mum says make sure you tuck your trousers into your socks.'

Suddenly there was another double cry from Cristoph and Byron. This time they were frozen, not running, and both pointing at the killer. The peregrine's belly was gold as the grass; it came in straight and fast and low and our cries made no impression on it. We watched it go all the way we had come in seconds, twist and disappear. We grinned at each other, speechless.

The walk took us in a slow curve to higher ground, then back towards the river and the camp. A line of quiet cattle followed a drier path. We stopped to watch them as they came, gently swaying along, attended by a very small boy who smiled shyly. The cattle's heads nodded gently as they walked and their hooves made hardly any noise on the path. The peace and ease of the beasts and their guardian were infectious. We stood still as they passed, as though witnessing something as old and simple as can be imagined, a pre-pasturalist scene, from a time when our forebears and their flocks were nomadic.

I set out to emulate our expedition at the same time next day. I hoped to see swallows, and Byron said there were hobbies, a fantastically manoeuvrable falcon which hunts swallows, which I longed to see – but my main purpose was to prove to myself that I could just go for a walk, somewhere in Africa, and not get eaten by anything. I circled villages, listened to children singing in a school, and eventually found myself a mile from the camp under a sky which had turned suddenly black.

'This place!' I muttered, hurrying back as fast as one can without running.

The storm drove spears of lightning into the land on the other side of the river. A girl was heading for cover 30 yards to my right and we slowed and smiled when we saw each other, pretending not to be scared, and there was a crack like cannon fire, an arrow of bright electricity, at least one of us shrieked and both fled. I paused again, later, just by the gateway to the camp. Just before dark is the time to see birds of prey, Byron had said, and he was right. I had watched a Fish Eagle earlier, up river, beating low and magnificently heavy not far away, but this was like glimpsing death.

Suddenly it was there, twisting and flickering in the dusk. Its wings were like sickled scimitars, black against the gathering dark, and it flew in a careless, chaotic way, like a malignant swallow, as if it did not care at all which way it fell next. Then it was gone.

'Bat Hawk!' said Byron enviously, back at the camp, as we pored over one of the bird books. I was delighted, and shaken. This thing eats bats! And yet, and yet, what I really wanted to see, the bird I had been looking for, was the Hobby. In Europe swallows do not have many predators. A Peregrine is fast enough to take them, but peregrines are more likely to hunt larger prey: pigeons are their game. The Sparrow-hawk is not fast enough, though they occasionally catch swallows by surprise, and the Goshawk is confined to forests: the one raptor which routinely hunts swallows, and even swifts, is the Hobby, migrating on the tail of the tide of swifts and swallows, from southern England and northern Europe down to the southern fringe of the Sahara. This beautiful falcon has long wings, a dark back, a handsome black moustache, and red feathers around its thighs, as if it were wearing scarlet shorts. But in Africa the swallow's problems multiply. As well as peregrines and migrating hobbies, there are African hobbies, eagle owls and bat hawks. Then, crossing the swallows' line of travel is the track of Eleonora's Falcon, an elegant, dark-backed bird which migrates between Madagascar and the Isles of Mogador off the Moroccan coast, as well as along the North African littoral into the eastern Mediterranean. Matching the speed of the sea wind, Eleonora's falcons hang together in hovering flocks, a curtain of hunter-killers, waiting for exhausted migrants. Add the rain, the deserts, road traffic, storms, human hunters and adverse winds, and the swallow's journey begins to resemble a giant chessboard, crowded with mortal dangers.

The immediate change that camp wrought in me was to do with fear of the threats that might attend my own journey. Knowledge of what snakes were in the reeds, of how animals behaved, of where the Dikongoro lived, of how quickly cerebral malaria can take you, an understanding of all the places in which death lurked and the myriad ways it could strike did not allay fear. What blunted it was seeing people disregard fears, laughing, as if they had accepted in advance whatever price Africa would ask of them.

In London the restaurants were booked solid. The London Eye was probably glowing pink; no doubt with a bit of planning and cash you could have rented a pod, complete with roses and champagne. Happy couples were kissing each other and more needful ones were texting. In Caprivi the bush telegraph had been busy with news of the evening's entertainment. Of particular interest, as far as the staff were concerned, was the question of who the Overlanders were being guided by.

'Who's coming?'

'Two Lucys!'

'Juicy and Posh?'

'No, just Juicy, and another one.'

In swallow terms, St Valentine's is a festival of human courtship, display, pair-bonding and copulation. Swallows compress into days a ritual which western human culture has stretched into years. Arriving at their European breeding sites, which – depending on their time of arrival and the ferocity of the competition – may or may not be where they themselves were born, males select and defend a territory and await the arrival of females. With the coming of the females the males begin their display flights; soaring and diving above their territory, singing and flaunting their tails. If and when a female is drawn to him, the male will land, fan his tail, give his enticement call and point out his suggested nest site with pecking gestures. If the female is unimpressed the male will need to suggest another site, quickly, or find another mate. On average a female will find a partner within three days of arriving; males are likely to court two or three females before wooing one. While just over four days is the average time it takes him, up to thirty-one days of trying have been recorded, before, finally, one bird succeeded.

If breeding is not a success they will split up – and it is the females who seem to make this decision, changing nest sites, and therefore mates. If breeding is a success and both partners survive to return the following year, they may breed again, travelling together over 12,000 miles. However, the perils of migration take a terrible toll of bonded couples: only around 10 per cent of pairs seem to manage to breed two years in a row, with very small numbers managing three or four years, and only one recorded pair breeding five years in a row.

And yet swallows are not always loyal. Both male and female swallows practise extra-pair copulation – they have affairs – and they are highly adept at it: between a third and a half of all nests contain a chick fathered by another male. For reasons which are unclear the most attractive male swallows (with the longest tails, in the best condition, which have the least difficulty finding mates) make the poorest partners and fathers, contributing the least to feeding and rearing chicks. Just as many males of both species father children with different females, so female humans and female swallows sometimes give up their offspring to other females – known as 'fostering', in human terms and 'egg-dumping' in swallows. The difference is that human females know they are raising someone else's child, while a female swallow who suspects the egg in her nest is not hers will push it over the edge. Infanticide is common in swallows. Males will kill chicks in poorly defended nests, in the – usually successful – hope of mating with their mother.

What swallows 'want', or, at least, their goal as suggested by their behaviour, is clear: to have as many offspring as possible. The aspiration of humans as suggested by our behaviour is more opaque, but perhaps only slightly, at least on Valentine's night.

By eight-thirty Juicy Lucy was dancing on the bar in a skintight 'leopard-skin', entertaining her troops, an overlanders' truck. By ten we were all dancing on the bar. Mark and Margie had passed by, Margie smiling, Mark trying hard not to, clearly thinking Thank God the rest of the camp is empty and there is no one living directly opposite us.

'Don't you know what a springbok does?'

'NO!'

(I was clinging to the best seat in the house, hitting various bottles, and there was every chance of dancing.)

A springbok – a small South African antelope, usually, but in this case the participant of a drinking game, as Lucy explained to the crowd – pogos to within a leap of the bar, exposes its white tail, looks right and left, shaking its tail, pogos the final leap to the bar, lowers its mouth (springbok have no hands) over its shot of Springbok (Amarula, amongst other things), and shoots it as best it can. Supported by the other Lucy, Lucy had no trouble persuading the group to drop their shorts and take part. Byron had done his bit, appearing in the leopard-skin in the first place, before ripping it off. Even a leopard-skin cannot match a good pair of pants. Lucy's persuasion of the two most senior members of her clan, Adventurous Travellers if I ever saw them, summed up the role of tour guide on an overland truck.

'YOU TWO CAN —— GET THE —— UP HERE —— *NOW!*

It was not clear to me if that was true though, because they were holding each other up, laughing helplessly. It was a night for laughing and dancing and I forget all the songs except one. The chorus sounded like:

'What if God was one of us . . . Just a slob like all of us . . . Just a some-some some-something ye-ah, ye-ah, God is Love, yeah, yeah God is Love . . .'

I could not help but look around and ask *What if God were one of us?*

And the answer was standing there, half-smiling, not quite wide-eyed, not at all drunk, despite my efforts, and beyond bemused: he looked hornswoggled. Earlier he had muttered: 'There are two people – over there. I think they are women!' and indicated the sundeck which overlooked the floating platform around the swimming cage, the dark Kovango and the water monitors' reeds.

Everyone called him something else but he told me his name was Joseph: he was the brother of the day barman (of the same name), filling in for the regular night man, who was off, and being given his first taste of the Dionysian side of the camp. The casual, raucous and extraordinarily permissive antics of the white world had perhaps not offended so much as shaken him: he smiled tentatively, nervously, as if witnessing something forbidden but unexpectedly funny.

The next morning we said our farewells and I was given one pearl of advice for the road to Zambia.

'Don't hit an elephant.'

'What?'

'Seriously, don't hit an elephant. You will not come off best.'

CHAPTER 3

Somewhere Like Zambia

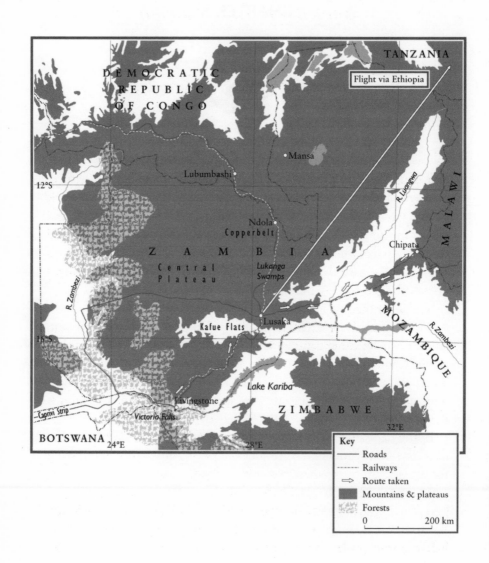

Somewhere Like Zambia

In certain parts of Africa, in the game parks where there are still large numbers of wild animals, a good way of finding swallows is first to find an elephant. Ideally, several elephants, in grassland, and on the move. As they go, swaying along, in the gentle, heavy-hipped way they have, like fat ladies lost in thought, elephants will disturb and put up a great many small insects from the grass. The swallows know this, and so they follow the beasts, and hunt.

'Come on then you elephantine chickens, come out and fight.'

The endless DANGER! ELEPHANTS signs had worn me out. Another storm came up with a fantastic bang, lightning landing like artillery, much too close. The road was dead straight and dead empty. There were only tall trees and occasionally clearings on either side. This part of the strip is a game park and the wartime convoy habit is dying hard.

Byron had said there were hobbies along here, loads of hobbies, but of course until you have seen a bird once it is doubly difficult to see it for the first time. I wished Byron was there.

The Mousebird beeped disconsolately, I swore at her, changed gear, and began singing 'Fairytale of New York' to keep my spirits up. All my self-containment was drained: it was a lonely business, leaving half a dozen people who had become fast friends, and taking the wilderness in exchange. I muttered to myself, scolded the car and sang the song's tune again, changing the words. '"'Twas the day after valentines, in Caprivi, and the signs said mind el-ephants, but there weren't any anyway" . . . Holy mother!' There were elephants. Six young bulls, right next to the car. They were standing under a tree, sheltering from the storm and all but winking as we sailed past; I saw their eyes; one elephant looked right at me, through the windscreen. Its expression was amused and happy: that elephant was eating, chewing ruminatively, pressed up against its fellows as they waited for the rain to pass.

I met another, later. He stood in the road just long enough to stop us, a quarter of a mile short, before ambling off to the south. I craned to see him as we passed his spot, but though the cover there was not thick he was quite gone.

I spent the night in Divundu, a steamy dump of a border town with persistent and hungry mosquitoes. The car had a wash in heavy rain and I shivered in a sort of caravan that reeked of damp and rot. I cleared out the Mousebird, brusquely. I knew no good way to leave a beloved. Someone was making a great pilgrimage up from Windhoek to take her home but I felt jealous rather than bad about it. Life would be lonelier without her. The way our possessions reflect our personalities and pasts back at us is an effective screen, a defence against the indifference of existence. We were a well-adjusted little unit, and I would have felt better about going into Zambia with company, albeit inanimate. I did not know anything about Zambia, except that it was one of the world's poorer countries.

It was a clear hot morning. Three irritated young women, and a mother and child, distressed because they were leaving Dad in Namibia, were waiting for a ride that had not shown. They allowed me to join them. When the driver turned up, the voluminous boot of his van was stuffed with ill-concealed cans of contraband petrol, which was cheaper in Namibia than Zambia. He was working as a white-water rafter at Victoria Falls, he said, and smoking a lot of dope, I judged. He got us through the border very quickly, the officers remarking that I was lucky I had got my visa in London as

Zambia had just slapped a stiff, hard-currency price-increase on its rate to Brits.

Zambia did not look so different from Caprivi, except that there were more undulations in the road, and everything was broken, except the airport. We went there first, to drop off the mother and child. In Livingstone signs, buildings, roads, kerbs and pavements were all broken. There was a tatty, ragged, transitory air to the place, as if people were here despite themselves: the tourists for the Falls, and the migrants for refuge from Robert Mugabe. Half the town's prostitutes, according to BBC news, were Zimbabwean. There were stories of trucks backed up for miles on the other side of the border, with nothing moving between them but girls.

The pool at the Jolly Boys backpackers' hostel was surrounded by miserable-looking girls; young Scandinavian teachers. Jollity was in short supply, thanks perhaps to an absence of boys, and also to the vast amount of kwacha required to buy a drink. There are no coins in Zambia. This means that everyone always owes someone change, and the smaller notes in unrippable and often soiled plastic are gold dust. Because computers had not yet taken over, barmen spent their shifts with furrowed brows making pencil entries in ledgers and juggling debts, loans, orders and 1,000 kwacha notes.

The Scandinavians wafted in and out of their dormitories, clutching passports and cash like virginities. They were stalked, hopelessly, by older and drunken Brits, South Africans and one or two white Zambians, who held and held forth views on racial and gender politics and the state of Africa, particularly Zambia, which no self-respecting liberal could tolerate for a second. I nodded, smiled, said please and thank you and wondered where the swallows were.

'We've got fuckin' hundreds of them. There's been a German couple staying with my mate who are doing research on them. I'll call him, he's got a microlight. I'll call him. Hey, Steve, there's a guy here who's into swallows. Yeah, swallows. Hey, can I give him your number . . . ?'

'Where does your friend live?'

'You can't miss it. North of Lusaka, first left.'

His name was Peter. We straggled out into the darkness to look for food. Pot-holes lurked like crocodiles in broad reaches of darkness. Peter soon decided that the bars were too empty, the women too proper, and he was not hungry. Like everyone else, he was counting his kwacha.

'I run a trucking firm. I've got one truck here, stuck in the mud, and I've got another the driver is saying is broken down. They're probably trying to sell the load right now. I tell you, this place, it's fucked.'

I decided to leave Peter's friends and their swallows to their own devices.

To see the wonder of the world that is Victoria Falls you sign in, hand over dollars and pass an ugly statue of Dr Livingstone apparently hewn from solid copper. You decide whether or not you need a raincoat and boots, and then follow one of the little paths towards the noise.

Perhaps Victoria Falls is the ultimate triumph of the British Empire in Africa; you are told its other names and they are all older, truer, sound more beautiful, say more and show more, but Victoria Falls is as mighty as English itself and as surreal as the great White Empress hurling herself off the greatest cliff on the continent and plunging to her death with all her petticoats billowing. The locals call it Mosi-Oa-Tunya: the smoke that thunders.

A man sold me a statue. She was beautiful: the first Queen of Victoria Falls, he said. I asked her name. Mukuni, he said, named after her people.

'So they should be called the Mukuni Falls!' I cried.

'Yes!'

In fact he named her after the first Chief Mukuni who came down from Congo in the eighteenth century, conquered the Baleya people (who were originally of the Rozwi culture in what is now Zimbabwe) and founded Mukuni village, which is to this day presumed to have been the largest village in the area when Dr Livingstone arrived. The rest is history, or rather politics, as there is still a Chief Mukuni of

Mukuni and you may visit him there and mix with his people, the Toka-Leya, many of whom are artists.

She was a tall, slender woman, and weighed less than 100 grammes. Her people call the falls Shungu Mtitima and perhaps her name was actually Bedyango, the high priestess of Mukuni, who has final say on who will be the next chief. But this was a wise man, so he gave her to me in return for a little understanding, not too much, and I carried her from then on, neatly swaddled in the middle of my emergency toilet roll, believing her to be the true queen.

I began the business of organising my next transport. Hiring a car took a day and a half, a small fortune and multiple trips to the airport, half of them through magnificent downpours. The rain was always preceded by a thick green-red smell, which seemed to steam as if from cracks in the earth; then the sky turned grey-black, sometimes shot through with silver-white sun lances, and it would seem even hotter, under the cloud, and then the rain would come.

I was not much enamoured of the car. White with a fake velour-like interior, it was made God Knows Where, but was unmistakeably Japanese. It was all electric. A boot lid that would cut your fingers off if you were careless, automatic transmission, quite a lot of oomph and no doubt thirsty, it had Grande or Classic or something in curly-wurly writing near the bank of its red and gold tail-lights. Power steering, with a ride like a feather mattress sliding down a spiral staircase, sod-all clearance and, it felt, twice as wide as the Mousebird (God help us if Livingstone's pot-holes were anything to go by) there could be only one name for it: the Pimp.

The traveller showed up at the hostel around supper time. He was so wiry and quiet he was almost difficult to see; he achieved a beer in seconds and eased himself onto the list for the cook's supper, though in theory it had closed a while before. Roan is a chef, a snowboarder, a Kiwi, about twenty-four, and he was coming to the end of three months on the road: Tanzania, Kenya ('It's a great time to go to Kenya,' he said, phlegmatically: the country was in flames and the

whole world was wondering if either Kibaki or Odinga could see a way
out), Malawi, Zambia, and, next, Caprivi. He was on his way to
Canada, via South Africa.

'Man, I'd love to see you in three months' time!' he laughed. 'You'll
be thin, covered in shit, you won't care what you eat, you'll be
obsessed with your ass – it gets really out of hand, like someone says
woah! I just did a beauty – like – you can go and see it if you want, it's
still there! And you're like no, really . . . And you'll be so at ease with
your body. I am. I'd tell anyone anything now.'

We played pool and drank. Roan showed me how you find south
using the Southern Cross.

'You take the long axis of the cross, there, and carry it on: that's one
line. And see below it, those two bright stars? They're the pointers. So
you draw a line bisecting a line between them, and carry it on until you
meet the line from the cross. Below that is south.'

In the end we did an exchange. Roan got my still almost pristine
little guide to Namibia and South Africa, and he was pleased to unload
a battered *Africa* on to me, which I eviscerated with a steak knife. The
whole of East Africa from Egypt to South Africa went, except for
Ethiopia, and I also saved the first or last pages of countries like
Zimbabwe and Mozambique which have borders with countries
through which these north-west-bound swallows approximately pass.

The plan was to drive north-east to Lusaka, Zambia's capital,
break, drive north through the Copper Belt, get as close as possible to
the Democratic Republic of Congo, peer at it over the border and
attempt to answer the following questions.

Could one obtain a visa for the country at the border? If not, was it
possible to fly direct from Zambia to Kinshasa in the DRC without a
visa for the DRC (starting in London, you have to begin looking for
DRC visas in Brussels, and I had not) on the basis that one was really
going to Brazzaville in the Republic of Congo, just a short boat ride
from Kinshasa across the mightiest river in Africa?

Should any of these be possible, I would still be sticking to the
supposed flight plan of *Hirundo rustica*. However, it seemed unlikely.
Tim Butcher, author of *Blood River*, a recent book about the DRC,

had set his heart and years of preparation on coming through Congo in one piece. The writer of an article for the *New York Times* took a train which sounded fantastic, but she was there for a month while I had a week at most. One blogger wrote an hilarious piece about obtaining a DRC visa from Zambia's northern border, which involved days and days and days of the dullest work there is: sitting still in great heat, waiting.

Which would leave, I feared, one option: Brazzaville via Addis Ababa, courtesy of Ethiopian Airlines, but I did not have to face it yet. All around, thousands of swallows were either preparing for it or beginning to migrate, or were already far over the northern horizon, on their way.

It was eight-thirty in the morning and I was going too. The hitcher wore a pale grey suit and smart shoes.

'Where are you going?'

'Hello.'

'Sorry – hello! Would you like a lift?'

He climbed in.

'I am going just into town.'

'Right!'

We set off. Automatic transmission and assisted steering do have points in their favour; the acceleration was good and unlike the Mousebird, the Pimp did not make any audible objections.

'So what do you do?'

'I am a person seeking employment.'

Fingers dive hopefully into a sheaf of papers in a folder. A CV tries to emerge, but I prevent it.

When I dropped him off the person seeking employment gave me a wave and a brave attempt at a smile. Zambia's unemployment rate is estimated at 50 per cent by the UN; life expectancy is less than forty. I was an old man, suddenly, in Zambia. The person seeking employment was not much younger than I, and I did not hold out much hope for him in Livingstone. Tourism has declined in recent years and the most recent employers to open offices in town were western NGOs and charities. I saw hundreds of people dressed for work on Africa's roads, but from then on I wondered how many were seeking employment.

I fell in love with Zambia just north of Livingstone. It was going to be an 800-kilometre day, the sun was shining, the tarmac patchy and popular with trucks, and there was a roadblock. The policewoman wore a smart pale brown uniform with a large peaked cap which made her look to me like a smiling summer version of an officer in the Coldstream Guards.

'May I see your passport please,' she said. 'Where are you from?'

'Britain!' I said. Her smile was infectious.

'This is your car?'

'No, I hired it.'

'Well. We are very pleased to see you!'

'Thank you, madam! I am very pleased to see you too!'

Wow, I thought. What a day.

A hundred kilometres later I had almost had it with Zambia. The road was dreadful. Only the locals, the daily locals, one suspects, could travel it at anything like speed. I tried hitching along behind one, a white bakkie with two grinning and carelessly chatting occupants. Although I could catch them on the rare good bits they vanished every time we hit the rough. I cannot have kept up with them for more than 10 kilometres. Doubly frustrating was the rolling green beauty of the country; you wanted to keep stopping but you could not: it would only confuse the slaloming trucks, snake-dancing ahead, headlights winking through the dust.

Then it changed: the miracle of a good road! I hardly dared celebrate for a while, but the way to Lusaka was clear. There was a rhythm in the land. It rose and fell in undulations, sometimes gentle, sometimes steep. The villages and towns generally sat on the north or east side of the streams and rivers, so you would ride a crest which dipped down to a bridge, then the road would rise again, through the conurbation. And there were swallows. I pulled the Pimp over and climbed out to say hello to a group of about a dozen. They were sitting on a telephone wire, it now being mid-morning. Either they had spent the night there, and were having a lie-in, diving off now and then to snack, or they were pausing, having been travelling since first light. Behind me there was a straggle of wire and a patch of marsh, picked over by egrets: a good place to hunt flies.

A man on a bicycle, coming up out of the next town, bobbed his head at me.

'What do you call those birds?' I asked him.

'Those?'

'Yes!'

'Ah, I am not sure . . .'

'Swallows?'

'Yes!'

'And they come through here?'

'Yes! With the rain. I don't really know their names.'

'OK, no – Thank you!'

'You are very welcome.'

'Thank you!'

So many conversations went like this. It was rare to find ordinary people who paid much attention to the birds – to any birds. Perhaps out in the country, away from the roads and towns, there were older people who noticed their comings and goings, but in Zambia there were few older people. Perhaps amateur naturalism, in the way it is practised in Europe, is a luxury afforded by security, wealth and leisure.

The swallows were a mixed crew; half moulted, half not: their tail streamers, the mark of a fully mature, fully moulted bird, were still very short. They are known to travel in loose groups but it is not certain if they journey with their mates, children or neighbours, or whether they simply join travellers they meet along the way. It is thought they set out from the north with family and associates, as long as the children are ready to go when the adults are. But in the case of late broods the parents will go first, the young following as best they can.

Long before we reached Lusaka, having been run off the road by a bus which chose to overtake on a blind bend, where it narrowly missed a vehicle coming the other way, I concluded that Zambian drivers are optimists. I lost count of the number of times some optimist, having just not quite killed either of us, sailed by with a blinding grin which seemed to affirm that God does indeed walk the earth and devotes His time to protecting all who love a summer day and can think of no better way of spending it than tootling through Zambia rejoicing in the thrills of the road.

AIDS was everywhere, according to the signs. A hundred variations of the message adorned a hundred boards. The bowed red ribbon, the pictures of condoms and the slogans were both ubiquitous and charmingly various, according to different painters' hands, but if the advice was pretty standard – get tested, use protection, be faithful to your lover – the strategy was remarkable. The scourge was made an unimpeachable force for National Unity:

TOGETHER WE CAN BEAT IT!

The Pimp and I had our own encounter with national health policy. At first, the queue of vehicles and the roadblock which extended beyond the road into the undergrowth and the armed men milling around made me fear it was something else. But it turned out that the line of Zambian disease control was just a little more thorough than the Namibian version. The Pimp had its tyres sprayed and I was encouraged to submit my hands to a nameless chemical, then allowed to rinse them under running water.

There is a long, long descent to Lusaka. The day was beginning to end when we arrived at the top. Lots of road signs had warned that we were coming to a serious hill, then the road tilted and we were on our way down. Cars and bakkies weaved through a rolling diagonal of trucks. Some were battering their way up, fighting the weight of their loads with all their might. Others were bumbling down, rocking and rattling. Some, on the ascent, were barely moving, grinding out clouds of black smoke. Others had stopped; one or two seemed to have died; bits were missing and their bodies were abandoned. Not long afterwards strange shapes appeared on the plain. They looked like weird beasts, monopods with odd-shaped heads. They resolved themselves into the concrete towers of Lusaka's main street: Cairo Road. The next morning I walked it.

Cairo Road at dawn is deserted. The trees down the middle keep the silence of night, while the pure plateau skies stream silver and rose. In an alcove, by a bank, a man wearing a hard hat returns the front legs of his chair to the ground and bids 'Good morning!' On the step beside him a green cicada twitches. In a gum tree on Cha cha cha Road pied crows quarrel and throw down twigs on parked cars. Lusaka's mosquitoes rest, fat with blood. The females are famous for their cerebral malaria: at the bar last night we slapped and splatted them, passed around sprays, lit endless cigarettes and called for more tonic to go with the gin.

The commute begins; hundreds come on foot across the railway lines, some wait for slow trains; all the rest jam Cairo Road. Yesterday evening a gang of road-builders came in, clinging to a bakkie, hanging off its sides and hooting with excitement as their driver skimmed them into the heart of the traffic. A policeman raised his baton like a sigh.

Breakfast is served in the Lusaka Hotel. Chinese businessmen scan the papers. Something is happening to the copper again: Chinese companies hold Zambia's wealth, ransomed to contracts, and Zambia has been protesting. Now Chinese spokesmen are offering cash back and undertaking to build clinics and schools. Zambia has seen all this before. The flags outside the Bank of Zambia are motionless. One is pinned above the bar of another hotel: the green is the green of Zambia; the red is the red of the blood of our fight. The black stripe is us, our people, the gold stripe is the copper, our wealth. And the eagle is our freedom.

The man beside me is hurrying his breakfast; he does not have much time; his contact will be here soon. Behind him a group of young Canadians are being given presentations on what to expect when they reach their postings. Often, they are told, impossible promises will be made. They are confused.

'But what do you do then?'

'Well, what do you do?' returns their trainer. 'That's kinda what we're here to figure out.'

On the tables out of the shade the scrambled eggs are brighter than sunlight and the tomatoes are flame. The hurried man accepts some of my toast. A conversation starts. He is a miner, and a mineral trader. He is carrying stones. 'May I see?' I ask, eagerly. Hesitantly, he displays his cargo. He keeps some of it in his sponge-bag which is never far from his hands. His fingers are cracked and flattened, the brown skin lined red. They have spent years digging Zambia's red earth.

'These are amethysts,' he says quietly. There is an opportunity here, suddenly. He produces his identity card, the plastic bent by the curve of his buttock.

'I am the trusted man,' he says. 'I am going to Livingstone to meet three South Africans. They will give me $2,000 for everything.'

Not just for the amethyst. The precious tissue paper is compacted in his pocket, but it unfolds artfully to reveal a shining little city of green stones. Now I have become a buyer. I choose instantly and offer. The trusted man pleads. I double the offer; two pieces of paper for one small stone with light green fire in its cloudy eye. We squeeze hands. I return to my room and stare at my stone. It is either beryl, which is pretty – and this stone certainly is pretty – but relatively worthless, and I have just encouraged the trusted man to rip me off, or it is emerald, in which case I have just got the most amazing deal, and a treasure. I feel like a gem smuggler, as I wrap it carefully and hide it in my rucksack. For all I know, I am.

Rain comes that night, with sheet lightning. It thrashes the corrugated roofs of Lusaka; even the dead tired and the dead drunk are shaken awake by detonations of thunder which seem to erupt from the ground. In a dark dawn that is half rain, half mist, we set out to drive the Great East Road. It is easy to find but hard to see. Only 20 kilometres along it we stop for something, and the Pimp will not start again, will barely blink a dashboard light: after half an hour we discover one of the leads to the battery has worked loose. Kristoffer grins at his sister; she laughs at another one of his amazing collection of younger-brother smiles. They were supposed to be taking the bus. Kris is working for a German NGO and Katherina is on holiday from her studies and her job with the State Prosecutor in Berlin. She has been going back to their childhood: they lived in Lusaka when they were small and she remembers it. Last night he went to bed in good time while she stayed at the bar. He got up to find they were now not taking the bus: Katy has met some Welsh (?) guy, he has a car, and has volunteered to drive them to Chipata, where they plan to attend the Ncwala, a tribal festival in the far east of the country.

Kris cannot decide if the guy is crazy or what; it was all rather confused, leaving the hotel in the rain and darkness; at first he thought the lift was just as far as the bus station. But now they are on the Great East Road. So he questions the driver gently, chats with his sister in German, decides that the situation is quite acceptable, and after a while, falls asleep.

We stop in a village.

'African Fanta is the best,' Katy declares. Their mother is English, their father German, and Berlin is home. So though they are both German they can be quite English as well. The sky has turned to a hot grey-white and the sun is directly overhead. I look up at a group of swallows. The light is peculiar, melted pewter and there is a penumbra around the sun. The road unravels in dark green coils down to the bridge, and the river.

The Luangwa is inadequately described by its designation a tributary of the Zambezi. The river is huge, a great splay of silvered-brown deep below us, and there is a checkpoint on the bridge. The guards smile and wave us through. There is a lot of traffic on the bridge today, several cars like ours every hour, as well as the usual trucks: it is the Ncwala weekend. Downstream, tiny on the great waters, a single canoe carries a fisherman. We seem very far from everything, as if the great river has carried perspective out of the forest and away. We do not know how many hours we have before us. The bus takes most of the day.

Broken into sections the Great East Road is not so hard. You climb out of the Luangwa valley and soon come to the land of the humpy-tumps, where the road switches and snakes, rises and falls, through round green hills, covered with forest. Zambia sits on the Central African Plateau. Most of the country is High Plateau (over 1,000 metres) and on the Great East Road you do feel up in the air: it is like piloting a small plane around bubbles of cloud. Then you are brought down by the pot-holes. There are so many different kinds you soon think of nothing else. The worst are the solitary ambushers, a sudden trap in a piece of good tarmac. Depending on the slope of the road they can be invisible until you are upon them. The most common are those which have allies strung right across the road. You have a choice between veering right, sticking left and zig-zagging furiously or hauling the vehicle clean off the carriageway and rattling along one of the dirt side-tracks, hard up against the bush. In some places, where the forests press in, in river valleys, the whole road has been washed away. Here there are diversions, and you bump over raw earth, around men and bulldozers. When you misjudge things you cry, 'Pot-hole!' or 'Hold on!' and your passengers brace for the crunch.

'Sorry . . .'

Katy and Kris became accustomed to the sudden decelerations and side-to-side swinging, and able to doze peacefully. They took turns at keeping me company, feeding me drinks and quiet conversation.

Children who live beside pot-holed stretches have developed a relationship with the little pits. In some places they work at filling them in with earth; in others they mark them with fronds. They operate a tax system for this service, smiling and extending hands for tips. Away from a village, twigs or leaves on the road mean hazard ahead: normally a broken-down truck.

'Swallows!' I say, again, pointing.

'Ah yes,' Kris nods, loyally.

They were going north; I saw a hundred that day, all heading in the same direction. The migration was underway.

The road climbed up to a plateau. The country looked like ranch-land; there were many goats and cattle here and there. We stopped for lunch at a lodge where they have a puff-adder problem. I slept for a few minutes and then we moved off again, down a wonderfully bad stretch where water and trucks had almost rubbed out the tarmac. There was an hour of bad road, then an hour of good, and then the country changed again. Suddenly there were wide gold-green valleys and proud hills. It looked like the dreaming background Renaissance painters made of Italy. And now there were people, people every-where. We had found the Ncwala.

There were trucks full of people and lines of pedestrians. Many of the men carried sticks, and wore leopard-print hats or girdles. A drunk riding a bicycle and waving a stick came towards us, veered away and toppled over, grinning confusedly as the bushes came up to catch him.

We rolled into the little town of Chipata; there were high eucalyptus trees and low buildings, side roads made of rain-ploughed earth and a man with a huge burning cone of marijuana was passed out in the car park of the police station. Outside the supermarket we met our connection: she was small, auburn-haired, had bright dark eyes, and dynamism seemed to spark off her. In my mind she became the Red Torpedo because there was no problem she could not sink. She was American and she was with the Peace Corps. Through her, and the Peace Corps, we whirled into the spirit of Chipata's biggest weekend.

'So, Kate, when you contacted me I thought it was going to be impossible but I have got you a room. It's quite a good place. There are three of you now? Oh, no problem, I'm sure they can put another mattress in. OK, so they have Ncwala every year. It's three days of dancing basically, and so so much drinking, and then they kill a bull and the chief drinks its blood. The President of Zambia is coming and everyone's really excited because there's a rumour going around that Prince Charles is going to show too – like Yeah, right! I'm not so into the big bull killing because I've seen it before and there are so so many drunk people. I thought tomorrow if you'd like we could go and see my friend in his village; it's about 20 kilometres or something and the road isn't good but it's incredibly beautiful, you can see Malawi and Mozambique. Ncwala's actually not a very old festival; it's supposed to celebrate the harvest but the story is that the people here came from South Africa escaping from Chaka and this is where they stopped.'

'Where do you live?'

'In my village – it's 30 kilometres away. Huh? I ride my bike. If I have to go to Lusaka I hitch. Huh? No, it's fine, I speak Chewa – my people are Chewa. Is your room all right? Sorry sorry sorry, one of you is on the floor.'

'Sorry sorry sorry?'

'Yeah, you've heard people saying that, right? It's beautiful actually, like if you hurt yourself I say Sorry sorry sorry to you, to make it better. Look are you guys OK? Do you want to come to the Peace Corps house later? We'll start there and then go out. Mumpy's playing – you don't know Mumpy? Oh God, everyone loves Mumpy. She had a big hit . . .'

In the Peace Corps house I understood the lump of fresh turd on the loo seat in our hotel. There was a sign on the wall, showing a man squatting with his feet on the seat, with a cross through it, and another showing him sitting, with a big tick. 'If it's yellow let it mellow, if it's brown flush it down,' it admonished.

The Peace Corps house was like a self-regulating university with anyone's guests free to come and go, and an invisible, absent authority who was said to be principally concerned that we did not upset the neighbours. The Peace Corps included a man who had given up his job in a Washington think-tank; Mark, who was worried by his lack of qualifications – his elders counselled him, over cane spirit and a violent orange concentrate – and J, who was not worried by anything at all. He had long dreds and a slow, wide smile.

The Red Torpedo talked of his doings with something like awe.

'J is amazing; the things he is doing in his village are wonderful. He's made a field and he's farming. The people think he's incredible. He is so young, though. Like, he kinda disapproves of anyone who does not smoke dope.'

'What are you going to do, J?'

'I'm going to farm.'

'Where?'

He thought for a second and smiled. 'Somewhere like Zambia,' he said.

The Red Torpedo is a trained agronomist. 'I tried to teach them about seeds and things,' she said, of the villagers she lived with, 'but they just weren't interested, really. Like, some girl telling them how to farm? Forget it. But my biggest success is self-defence classes for the women.'

'What are you teaching them?'

'It's kinda Karate.'

'Oh, you know Karate?'

'No it's all based on the Karate Kid.'

'What!'

'Yeah, you know, wax on, wax off? The women and kids love it and the men were all really scared of them at first. God, though, the things I've seen. Like I went to a seminar on drug education and there was this expert telling people how to get their kids off dope. Put them into a bed and then somehow raise their body temperature until they begin to foam at the mouth! I was like, er, excuse me . . .'

We were going to Blue Gums bar. The Red Torpedo told us that everything we carried would be stolen, and that we would be assailed by Hoolays.

'Wholays?'

'Hoolays. Hookers.'

An hilarious child was to be our taxi driver. The car was as crocked as he was young.

'How much have you drunk?' I demanded, severely.

'Nothing for half an hour!' he cried, indignant. He abandoned the wheel to answer a text message. We veered.

'Hey, we nearly crashed then! But you took the wheel! That was good!'

'Thanks.'

Blue Gums is approached down a landslide pretending to be a road. A white bakkie with a familiar logo bullied us into a pit.

'Fucking Celtel!' screamed the driver; they were sponsoring the Ncwala; they were everywhere.

'That's all very well – who provided you with that text message that just nearly killed us?'

'Ha ha – fucking Celtel!'

Blue Gums rocked with Lingala, Congolese music, mixed into the theme tunes of central and west Africa 2008: Akon, 'No-body wanna see us to-gether, but it don't ma-tter-no . . .' and Shakira: 'Oooh baby when you dance like that . . .'

We threw ourselves in, Katy grinning wickedly at her brother's restrained bopping.

'Come on, you Welsh sheep!' he yelled. I soon found myself royally beaten in a dance-off with Robert, a Zambian working with the Peace Corps.

We left Blue Gums before the dawn, riding on a bakkie filled with smuggled lager from Mozambique and something called Imperial Russian Vodka which was distilled in Maputo, Mozambique.

They were still dancing in Blue Gums two days later, when we returned to drop off a boy called David, after the Ncwala. He called me David too.

'Why have you guys been calling each other David?' the Red Torpedo giggled. 'His name's not David – neither is yours, is it?'

It did not matter. We agreed that he was going to be President of Zambia one day and I was going to be his chauffeur.

At the Ncwala we sidled into the stand at the centre of a huge circle of people, the main arena, where the remains of the bull were smouldering on a fire, and sat near the then incumbent President Mawanasa of Zambia. I tried to imagine an unknown Zambian turning up at Wimbledon and drifting into a seat near the British prime minister.

'Mzungu power' the Red Torpedo called it, with a dry laugh. Mzungu is the common and partly derogatory term for 'white'.

'David' briefed me on the various politicians as we did our best to be good ambassadors for Germany, America and Britain among all the chiefs and ministers. We sat up straight or surrendered our chairs to the chiefs' wives and children. We had missed the killing of the bull but we were in time to see the president solemnly accept a slice of its meat. Moments after swallowing it he left, with dozens of the other dignitaries, in a regal rally of pristine vehicles from which soldiers protruded, ever alert, weapons at the ready.

The dancing men shook the earth with their feet; the women provided chorus lines of swaying rhythm and chanted melody. Sometimes the dancers charged us, slaughtering our shadows with their spears and sticks. The crowd downed torrents of Chibuku Shake-shake beer or huddled under umbrellas bowed by the weight of the sun.

'There was a battle on the other side of the hill,' one of the seated men explained.

'They beat some British there, so they are also commemorating that!'

The Ncwala is promoted in Zambia as a significant national festival, a date in the cultural calendar, a vibrant contribution to its nationhood. Yet, in the midst of the crowds, the noise, the drunkenness and celebration, there was an absence, a bare space. It was impossible to know if knowing that the festival was a relatively recent fabrication, a tradition under construction, contributed to this empty feeling. Would it have had more force if it had taken place every year for centuries? Does the authenticity of culture intensify with age and repetition? Did the people celebrating the flight and eventual coming to rest of their ancestors feel richer, more anchored, more of a people for the fact of the celebration? As a Zambian event, attended by the president, did it contribute more than publicity to the essence of Zambia? Or reflect anything deeper than a young nation's desire to create ways of honouring a past that was older than the nation, older and more authentic than the colonial wheeling-dealing that created it, sixty years ago? Everyone I spoke to mentioned that the Ncwala was not very old, and in the hordes of drunk men wearing 'leopard-skin' and waving sticks I could detect nothing deeper, more moved or moving than a kind of tribal pride. The tribes, the Chewa, Bemba (everyone else made jokes about the Bemba) and Ngoni, whose festival this was, were the true cultures of this place, and the country. There was nothing older than their traditions, nothing stronger than their loyalties. The triumph of Zambian culture, such as it was, was that at the Ncwala there were several tribes present, and that they all acknowledged the president as their leader. Perhaps every nation is a constellation of different peoples, and perhaps the extent to which they allow themselves to be harnessed, on the one hand, and on the other the degree to which they feel themselves represented by the federation, the contemporary state, is a measure of the strength or weakness of a national culture. It was striking that the flag of Zambia was not prominent at the Ncwala – while the banners of Celtel were everywhere. The root of the Ncwala, the story of escape from a darker past into a comparatively free present, is also the story that Zambia tells itself, the red of the revolution in the colour-code of the flag that every child learns to read in school. Prosperity, or at least peace, out of struggle: it was the same story as Namibia's and South Africa's too. It seemed to beg a question of borders. Why, if all these countries are mixtures of tribes, recently created out of a colonised past, are there borders between them at all? A few miles from the town, those borders dissolved.

We followed a road out of Chipata, through maize fields past a chief's village, around and up until we came to a stream bed in deep shade; this is the smugglers' track. Men on bicycles and occasional bakkies bumped down towards us. The road climbs through thick woods, then levels. There is a village below. Here a friend of the Red Torpedo, who *is* called David, was posted by the Peace Corps. The villagers had made him a beautiful hut, strong and comfortable, enclosed for protection against snakes and goats. David's 'father' is a currency dealer, driving thousands of kilometres between Zambia and its neighbours: David says he is doing very well. His 'mother' is sitting on her step, sorting beans.

Below the village is the valley where I instantly longed to live. It was only a mile or so to the round hills on the other side. To the left the ground rose up to Malawi, to the right, with gentle bends, it meandered down to Mozambique. The valley was a rich collaboration of hot summer yellows; on the slopes of the hills was a parkland scattering of trees. Everything grows down there, David said: citrus, grapes, wheat, avocado, maize, tobacco – anything you cared to plant. It is all irrigated by streams running down from the hills. The valley does not seem to be anywhere. The names of the three countries that surround it sound like the abstractions they are.

We sat outside David's hut listening to stories.

'There was a man who needed to see the witch doctor about his problems. He went to the witch doctor but before he had even told him about his troubles the witch doctor asked him for money. The man said No way! You have not done anything for me yet! Why should I give you money for nothing? and he went back to his hut. He had not been there very long when he noticed something moving under the covers on his bed. He thought, Ah-ha, I am going to get that mouse this time. He fetched his machete (don't ask me why he wanted his machete to deal with this mouse), prepared to strike, and ripped the covers off his bed. There was a Spitting Cobra. He dropped his machete, grabbed his money, ran back to the witch doctor, gave him the money and said Here! Take whatever you want, but no more Spitting Cobras – please! The witch doctor took the money and the man never saw the snake again.'

You would not pay much heed to such a story, perhaps, if there were not a good population of Spitting Cobras, invisible but present throughout the green country all around you, and if every village did not have a witch doctor, and if the vast majority of the people you passed did not believe in their powers. The story seemed to illustrate how western secularism and Christianity had penetrated this part of Zambia: far enough to raise questions about the powers of the spirit world, but not strongly enough to overturn them.

*

'Look, swallows! What do you call them?' I asked Robert, the Red
Torpedo's friend, a Zambian who helped the Peace Corps, and hoped,
somehow, perhaps with the help of one of his American's friends, to
go to agricultural college. A group had appeared near the supermarket
on the edge of Chipata, where we were waiting for Kris. ('Quite
acceptable' was Kris's preferred verdict of approval, except for that
supermarket, with which he declared he was in love.)

'Nyankalema!' Robert said, 'We say it is the bird that never gets
tired.'

I could have hugged him. 'Nyankalema,' I repeated. 'The bird that
never gets tired – finally, someone who knows something about
swallows.'

Robert laughed and shook his head. The Mzungu and their
enthusiasms, you could see him thinking.

'You are Nyankalema!' he told me later, laughing, but it was not
true. We set out to drive the Great East Road again on nothing like
enough sleep. The Red Torpedo came with us, rather than hitch: she
was due to meet some new arrivals in Lusaka and induct them into the
ways of the Peace Corps.

'You get a tiny amount of dollars so you have no choice but to live
with your community, and what you do in your village is really up to
them and you.'

'Do you think you have been helpful?'

'Oh sure, but you know, it's really hard. Like the whole Aids
Awareness thing. What do you do if it's a Chief's right to have loads
of women, he's an alcoholic who doesn't use protection and everyone
puts up with him because, you know, men are men, he's the chief and
men just love to fuck?'

The road signs were just as straightforward.

HIGH SCHOOL STUDENTS

GET A'S

NOT AIDS

ZAMBIA NEEDS YOUR BRAINS

The Great East Road had grown longer. I became an automaton, a part of the car, fighting sleep, heat and the pot-holes. I know there were swallows because I said 'Swallows!' at regular intervals – even more frequently than I had on the way out.

By the time they reach latitude 12° south, where the plateaux of Zambia disappear into equatorial forests, swallows are either bearing north-east, via the Nile Valley, the Levant and the eastern Mediterranean into central Eurasia; or due north, towards the Sahara at its widest stretch; or north-west, towards West Africa and a narrower slice of the Sahara leading to the mouth of the Mediterranean, at the Straits of Gibraltar. My bird, if she lived, must be following the third route, as must many of northern and western Europe's swallows. For any that do not use the Rift Valley and the Nile there is no choice: they must cross the Congo.

In Lusaka I gave up the idea of going up to the border of the Democratic Republic of Congo. The younger brother of a man who runs trucks through Congo down to Zambia told me that it was perfectly possible to cross the country, of course. All you had to do was pay at every roadblock and pay again at the next. The only difference with the rest of the trucks' routes, he said, was that in the Democratic Republic of Congo you do not have the pleasure of bargaining. He did not think it would take me more than a couple of weeks to cross it, if I was lucky and generous with bribes, but he did not recommend it. I decided to fly to the Republic of Congo instead.

Lusaka to Brazzaville in the Republic of Congo is a distance of over 1,000 miles, north-west across the forest, but this was not a direct flight: instead it went via Addis Abba, in Ethiopia. We arrived in Addis at night and circled above the tumble of the city's lights, waiting for the stars of other Ethiopian Airways planes to land. Twelve hours in the city became compressed into two sessions in Bole Airport, a breakfast with superlative coffee, a ride through Addis' rush hour, and another take-off.

We flew back, south down the Great Rift to Kenyan airspace, then

west across Uganda, and the Democratic Republic of Congo. I
pressed my face to the window as we began our descent. I had heard
and read so much about that fat grey slug of river under greasy cloud.
The scale was difficult to grasp. There was forest, then cleared
ground, then Kinshasa: huge, with skyscrapers. There was the Pool, a
vast swelling of the river into something like a great lake. It narrowed
to a neck between Kinshasa and Brazzaville; as we banked over to the
west of the Pool I saw the waters narrow and change colour. Now they
were brown and streaked with cream foam, and there were rapids. I
did not understand it. From this height you should not be able to see
rapids, their strained narrowing, stretched water and shoaled waves,
but down there, there they were.

CHAPTER 4

Congo-Brazzaville: A Quiet Little Place

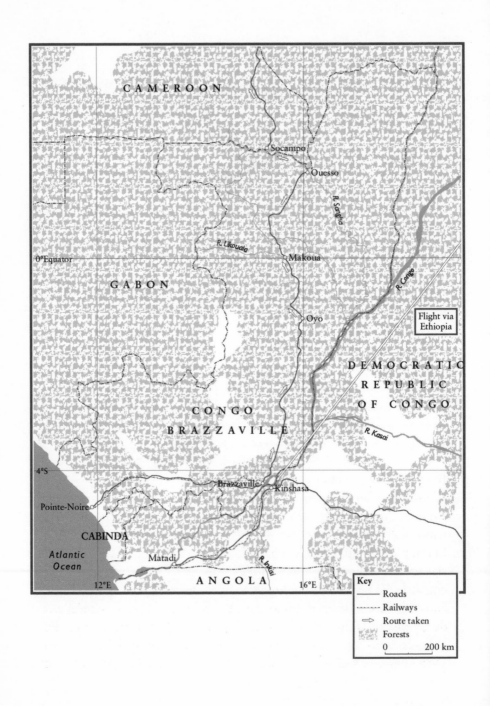

CAMEROON

Socampo

Ouesso

R. Sangha

R. Likouala

0°Equator

Makoua

GABON

R. Congo

Oyo

Flight via
Ethiopia

DEMOCRATIC

REPUBLIC

OF CONGO

CONGO

BRAZZAVILLE

R. Kasai

4°S

Brazzaville

Kinshasa

Pointe-Noire

CABINDA

Atlantic
Ocean

Matadi

R. Inkisi

12°E

ANGOLA

16°E

Key

—— Roads

·········· Railways

⇨ Route taken

Forests

0   200 km

Congo-Brazzaville: A Quiet Little Place

Six hundred years ago there was one kingdom of Kongo; today there are two Congos, according to the maps and flags. The Democratic Republic of Congo covers a million square miles of the heart of Africa, while the Republic of the Congo, otherwise known as Congo-Brazzaville, is a much smaller curl of territory on the north-west side of the river basin. The former was the personal fiefdom of Leopold II, King of the Belgians, claimed and seized for him by Henry Morton Stanley, one of the more remarkable and terrifying Welshmen ever to walk the earth. Ten million people died for Leopold's profit in the Congo Free State, a giant sort of labour camp which the monarch himself never visited.

On the other bank, the British having failed to exploit it, Congo-Brazzaville was colonised by France. While Leopold employed a Welshman who changed his name and pretended to be an American, France's pioneer was born of Italian descent in Brazil and became a naturalised French citizen: Pierre Savorgnan de Brazza. While Stanley deployed murder, coercion and torture, de Brazza loathed slavery and believed in the values of France: freedom, fraternity and equality. On behalf of his adopted country he used patience, dialogue and negotiation in his dealings with the people of Congo. Trade and development would enrich both France and Equatorial Africa, he believed; relations between their peoples could be humane and just. In 1880 he signed a treaty with King Iloy, and Congo west and north of the river became a French colony. France and Belgium shook hands over their maps at the Berlin conference of 1884. The rights to the rubber, ivory, wood and minerals of the Congo were divided to European satisfaction.

The consequence for the peoples of the Belgian and French territories was the same: around half the population of rubber-growing French Equatorial Africa died under the bullets and whips of their colonisers. Having been dismissed from his post as governor in 1889, de Brazza was wheeled out of retirement and commissioned to compose a report on the condition of French Congo in 1905. A cover-up was prepared for him, but it was inadequate. Horrified by the nation-scale abuses he had discovered and weakened by dysentery, de Brazza died on the way home. He was given a grand funeral in Paris. The French parliament voted that his report should be buried too.

The plane descended over a rainy land of rivers and tributary streams, small hills, houses, shacks and green rolling ground. Long since cleared of forest, the area around Brazzaville was described in the 'Lonely Planet' *Africa* guide as looking remarkably like Wales. I was not convinced: it was a yellower-grey. The centre of the city has one distinguishing feature from a distance, a round building like a length of pipe standing upright, slightly flared at each end. This is the Elf Tower, headquarters of what was the French state oil company, now subsumed in TotalFinaElf.

The three most striking travellers in the cabin were broad twenty-something Congolese in baseball caps, bright motorcycle jackets, gold chains and dark glasses. They chewed gum and lounged in their seats as we approached the runway. They were succeeding so thoroughly in looking like gangsters that they could only be Fashonistas, or Sapeurs, as they are called in Brazzaville, where style is very important. The rest of us were grey in comparison: African businessmen and Euro-American employees of one sort or another.

We landed in a rush past a Russian Anotov transport, old Boeings and smaller planes. The airport had a flaked, one-eyed look as though it had been recently shot up. In fact, the last time it was bombarded was 2003, when the former 'Cobra' (now government) forces of the former Marxist dictator (now president of the republic) Denis Sassou-Nguesso last defeated the 'Ninjas', led by Pastor Ntumi, who believes himself sent by God to liberate the Lari people of the south. Before that the airport, Brazzaville and much of the country had been fought over in 2002, 1999, 1998, 1997, 1994 and 1993, the last five conflicts being partly sponsored by the Elf oil company on behalf of its interests and those of its beneficiaries and associates in the French government at the time.

Sassou-Nguesso, a French-trained paratrooper and a northerner who needed plenty of guns and money to take power from his elected predecessor, Pascal Lissouba, was France's man. With vast villas in Paris, an unknown sum in offshore accounts and an estimated \$250 million leaking annually from the country's oil account (principally but not exclusively in the hands of Total) it seemed safe to assume he was still.

The human price for this status quo must seem very small by the historical standards of the Congo. In 1997 at least 10,000 people died and more than 800,000 were displaced. In 1998, during a Ninja counter-attack, a good proportion of Brazzaville's citizenry fled their homes. That year, on one notorious road, over a thousand women and girls were raped.

We had touched down, as far as I knew, into an uneasy peace. The Ninjas still held territory between Brazzaville and the sea, effectively cutting the capital off from the best deep-water port of the West African coast, the country's oil city, Pointe-Noire. They were said to be friendly now: you were supposed to be able to pass their roadblocks just as you would the government's – except, of course, anyone who could would surely fly.

As I walked across the tarmac, through dripping heat, I was sure of only two things about Congo. First, every account I had read promised I would meet the spirit world somewhere on my travels, Congo-Brazzaville being as much in thrall to ngangas (witch doctors), sorcerers and magic as it is to oil and guns. I was not looking forward to this. Second, according to my visa I had twelve days to make it to the country's northern border and cross into Cameroon.

The queue for the booth in which a man was stamping passports dwindled as local fixers and resident ex-pats appeared from behind the booth to pick out their associates. These were waved rapidly through. My turn came, the passport was scrutinised, stamped and handed back with a smile. There was a small, churned crowd of hopeful faces offering help with carrying the rucksack, then customs.

'Do you have anything for me to eat?'

The customs officer was a young woman in a brown uniform. Her gaze searched mine then flicked over my shoulder while her hand rifled the rucksack. She spoke urgently and furtively.

'Do you have something for me?'

Each time she repeated the question something else was pulled out of the sack.

'Do you have some money for something to eat?'

I surrendered two packets of duty-free cigarettes: the moment the first appeared, a second customs officer swooped with the same questions. They divided the cigarettes and let me go. In a *bureau-de-change* I changed euros to CFA francs: these are pegged to the euro at the rate of 655 CFA to 1, the exchange rate being guaranteed by France and underwritten by an agreement which puts 60 per cent of the CFA zone's reserves in a bank in Paris. Then a man with a small green Peugeot taxi whipped me into Brazzaville.

'What's it like here at the moment?'

'It's good.'

'Really?'

'*Oui!* We have security . . .'

'What's it like in Kinshasa?'

'Better! The son is better than the father!'

This referred to the presidency of Joseph Kabila, who took over from his father, Laurent-Desiré Kabila, in 2006. We laughed at the idea that the son could be better than the father:

'So often it is not the way!' the taxi driver said.

I checked into the best hotel I could find, planning to get my bearings and move downmarket tomorrow. A text message from London said an old friend had become a father. Messages from home came infrequently now. I toasted the baby, Oliver, in South African 'Flight of the Fish Eagle' brandy and went out to explore.

Tall palms line the river, a slow-moving rink stretching across to the port and the towers of Kinshasa. Large rafts of weed, water hyacinth and torn branches sail down the current. Poto Poto is a dense nest of life adjoining the city centre; its streets are swamps of brown mud and spume puddles which conceal pits and shell-holes, some deep enough to swallow a car. Poto Poto is principally made from corrugated iron. Workshops, garages, table-football bars, private dwellings, scrap-metal merchants and furniture-makers are tacked together, loud with saws and music, hammering, shouts and laughter. Smells of smoke, cooking meat, rotting garbage and coffee eddy among clouds of exhaust. People are friendly, curious and polite. Another cap-and-shades boy stops me for a while-away chat. Where am I from – what, Portugal?

'Non, Pays de Galles – un petit pays entre Irlande et Angleterre . . .'

To not be from somewhere simple is part trick, part blessing. By the time I have explained where Wales is, sometimes with a sketch map of the British Isles, the point of the question has been lost, the conversation has passed nationality and with it, perhaps, to a degree, preconception.

'It is a little country of mountains and rain, and lots of sheep. We are farmers.'

I produce the same line again down by the river, in a shack in the last of a line of shacks at the port of Brazzaville Beach. Disused railway lines lead there, past the walls and thick greenery protecting the Russian Embassy. The port is edgy. The soldiers look suspicious, a drunk man in uniform starts to demand money before swaying and giving it up with a whoop. Women are cooking fish on lines of griddles, the dust at their feet covered in entrails and blood. Children are smoking *'tabac congolais'* – marijuana – and grinning through the slats of the shack where we sit. On the water two ferries lashed together are docking: one threshes the river, carrying the other like a crippled sibling on its hip. Inside, flies crawl around the rims of the Primus bottles: the few of us who can afford to are drinking; the rest are sitting around, smoking. The men are arranged around the walls of the little room and our discussion is being followed with the earnestness of a court in session. My interlocutor is the biggest man in the room. His large head is minutely shaved and his brow is furrowed, his expression clouded with scepticism.

Pays de Galles does not satisfy him. He nails it to England, and England to America. Yes, I concede, politically we are Anglo-Saxons, but we are Europeans too, and Europeans are many tribes. Many Welsh, for example, say they are Celts, like the people of Brittany in France, or the Basques . . .

Anglophones, Francophones, vous êtes les mêmes . . . You are the same, he says.

Well, I counter, are you the same as the other Africans?

'*Non!*' He smiles slightly at last. 'We are Congolese!'

The observers start to laugh. Conversations resume.

'And what do you do?'

'I am with the port police.'

'You must be busy – there's a lot going on, isn't there?'

'Oh yes.'

Wilfred was off duty now, he explained, and going home to his wife and children. If I would like to come again tomorrow morning he would show me around.

In the middle of town a line of restaurants clung together under a white colonnade. One served kebabs or cakes over a counter, the next breakfast and lunch. 'They are all good,' I was told. 'If it's good and it works it is normally run by the Lebanese.'

From behind a pillar came American voices: two men in shades were finishing their meal. They were both broad-shouldered and wide-framed. One was bald and moustached; the other had an impressive beard. Apologising for interrupting them I asked if they knew anywhere which had a pool table. A pool table is a wonderful place to meet people, to find conversation and friends. Sure, they said, and gave me the address of a place, both restaurant and hotel. They were friendly and kind. They would be there later, they said – perhaps we would meet?

I moved out of my upmarket tower into the hotel they recommended, a long bungalow of rooms below a dining area, wound around the bottom of a tall building which was slowly being repaired. It was owned by Philippe, a laid-back young Frenchman who had cycled around the world, met his Vietnamese wife here, and settled. Members of her family worked in the kitchens, and the restaurant was deservedly popular, attracting the cast of a dozen novels. Among the regulars there was a Frenchman with a moustache so huge it looked like a disguise, and a Belgian with an impressive hat. They both looked as though they were in fancy dress, 'disguised' as spies. Passing trade included aid workers, diplomats, oil people and UN staff, European and African nationalities I could only guess at. At any mixed-sex table the women were always younger than the men. Beyond the buffet my new friends were eating with four others of their kind: big, broad American men.

'So what are you guys doing?'

'We're building the new American Embassy.'

'How's it going?'

'Real good. No problem.'

'Is it like the old one?'

'It's much better. A block back from the road, near the airport, plenty of space between the perimeter and the building, so if we have to we can pick 'em off as they come over the wall . . .'

The embassy-builders' lives fascinated me. They had all toured what seemed like dozens of countries, living in a travelling parallel world.

'What's the plan tonight?'

'You're coming out with us. We're gonna have a few drinks here, and then we're gonna go to a bar and get a few more, and then we're going out dancing, and you're gonna see more beautiful women who want you than you ever have in your life.'

We went to a dark cellar bar, air-conditioned, and drank while the clock turned towards night-club time. Some Congolese women turned up. We were all introduced with great courtesy by Jim, the extrovert king of the group, the man whom I had first spoken to. He had built embassies in more countries than I could count. Brazzaville was a hell of a lot quieter than some of the places he worked in, 'But it does have some good women!' he said.

His beautiful girlfriend laughed and kissed him. He was helping to put her child through school.

'It's only a small embassy,' said another man, whom I privately christened the Quiet American. 'What did you say you do?'

They were careful with me, saying as much as they could and automatically stopping short of a shared silence, a place like a wide, forbidden compound of things they could not say. There was one language we could all speak, as men far from home, and another only they spoke, which they would never use in public. They apologised for it with jokes and references.

'You better watch out!' Jim laughed. 'He's going to have to open a file on you now.'

I seemed to remember reading that Americans serving in the UK had to report encounters with the British public as alien contacts. I teased them about it. The Quiet American did not quite blush as he looked at the floor.

'How does Brazzaville compare with other places you have been?'

The Quiet American pursed his lips. 'Honestly? I would say this place is as dead as dead. There is nothing happening here at all.'

Somewhere out there, beyond the land, according to a paper I had read, two American warships were circling the continent, while diplomats tried to find a country prepared to host Africom – a military hub, a US equivalent of the French base in Djibouti. According to the article they were not having a great deal of luck, if the pronouncements of African leaders were true.

The embassy-builders said they did not know about that, but their construction would have lots of potential.

'There's only gonna be a few Marines stationed there but you should see it. There's bunks and showers and even TVs all ready. If they ever have to send a detachment here all they do is switch it all on.'

The night club divided into three strata. We stood at a bar scattered with other ex-pats; European men, late twenties to mid-sixties, lit by bright down-lights and multiplied in mirrors behind bottles. In the dark, against the opposite wall, were Congolese men. They sat in small groups and huddles, their eyes sweeping the bar and the spaces between us all, in which there were the girls. They outnumbered the men dancing three to one. If you met any of their eyes it was taken as an invitation.

'What do you think it's like, for those guys,' I asked one of the Americans, 'watching us? Don't you think they must hate us?'

'It must be weird for them,' said Dino, another of the embassy-builders. He was a broad Italian American who specialised in electronic security.

'That girl, she's beautiful,' he sighed. 'I'm gonna ask her out.'

That girl was beautiful. Anna spoke French, Dino English: I became a translator.

'Tell him I think he is very nice.'

'Tell her she is very pretty.'

'Ask him if he has a wife.'

'Tell her no, I don't! Is she married?'

'Ask him where he is from.'

'Tell her I'm American. Ask her where is she from?'

'Tell him I am from Kinshasa.'

'Ask her what she is doing here.'

'Tell him I am here for a rest! Ha ha ha! I have a hairdressing business.'

'Tell her she has beautiful skin.'

'Tell him I know he will have other girls.'

'Dino, she says she knows you will have other girls!'

'No way! That's not true. Ask her if she would like to go out with me tomorrow.'

'I am busy tomorrow but I will go out with him on Sunday.'

'Right, that's it! You guys understand each other just fine; I'm out.'

Wilfred was transformed. Now he was wearing a black T-shirt, with POLICE stamped on it in white, spotless black combat trousers, black combat boots, and a black automatic on his hip.

'Do you ever have to fire the gun?'

'Not yet.'

He led the way past the bars, through broken gates, to the port proper. Under the flat, hot sky a large market was in progress. Sacks of food were being unloaded from barges: a load of beans from Cameroon had just arrived. Spread out on the ground were fabrics, cassava, fish, chickens, vegetables, rice, spices, wood and a bundle of young crocodiles, alive, and all for sale. The crocodiles had their jaws lashed tight around wooden bits. They were not much under 5½ feet long and looked sunburned under their brown scales. Only their eyes moved, as if they were pretending to be a stack of planks.

'So what do you actually do here, Wilfred?'

'I make sure it is secure.'

We walked along to some warehouses where a group of birds were fluttering around iron beams. They were hirundines, but not, I was sure, Barn Swallows. We wandered back to the market, Wilfred pointing out the ferry to Kinshasa which was coming in sideways, stemming the current. Suddenly around us there was a chorus of whoops and shouts.

A thin man with bare feet in a ragged shirt and tattered shorts was running for his life.

'A thief,' said Wilfred, mildly, not moving. The man was tearing away from the crowd which was stilled, staring. Two or three men ran after him, sprinting as fast as they could. He ran first directly away from the river but quickly slowed on an unstable sandbank rising to an iron fence. The pursuers gained; one grabbed at him. He leapt away from the clutching hands and people in the crowd cried out as he changed direction, now coming down the bank. It had been apparent from the first instant that he could not escape: where could he run? The pursuers closed again. The fugitive stumbled, his foot slipped on sand and he crashed down, captors piling over and onto him. The crowd sighed. Men dragged the thin man to his feet and hustled him along, his head hanging and his steps uneven, weak now, as though all his strength had gone.

'What are they going to do with him?'

'They are taking him to the police post there,' said Wilfred. 'If we were not here they would kill him.'

Five minutes later, having taken his time passing between the knots of people in the market, Wilfred led me to see the man.

'What did he steal?'

'I don't know,' Wilfred said.

The man was sitting on the floor against the back wall of the police post. One of his knees was bleeding. A policeman sat between him and the door. The captive looked up. Wilfred surveyed him, then exchanged words in Lingala with the guard. There was no expression on the thin man's face but his eyes seemed to stare in spite of themselves. He was still panting slightly. We were all sweating.

'What will happen to him?'

'He will be fined, but if he does not have any money they will let him go later.'

It seemed I was expected to talk to the man.

'Are you all right?'

He nodded.

'Is your leg all right?'

He nodded again. There was nothing more to say, and the future seemed entirely uncertain. Would the man be beaten? Killed? Nothing seemed determined, likely or impossible. Wilfred and I returned to the bar, to drink and gaze at the river.

It was staggeringly hot in town later. A wind blew up clouds of dust from the roads which seemed to make the air hotter. The heat beat behind my eyes and in my temples; my legs shook like the thief's. You had to walk very slowly, or sit down.

I followed the sound of a celebration through deserted streets. In a forecourt behind railings in front of a government building people in bright clothes were singing and clapping their hands under a banner which said this was a celebration of the inauguration of new delegates – delegates for what it did not say. There were no bystanders; just women and men dancing and singing, in the heat, behind the railings. Despite their swaying and singing and the bright colours of the women's dresses there was a strained, isolated atmosphere, as if this was a staged rehearsal. I approached but then an army truck appeared, open-backed and stuffed with soldiers in purple berets. Their weapons protruded from the truck like spines: they did not carry their rifles at the vertical, but at the ready. They were not like any soldiers I had ever seen; neither cool professionals nor bored recruits; they were tense, tight-eyed, their faces blank; it was impossible to tell what they were thinking: pain at the heat, anger, anticipation of trouble – some sort of attack? I fled.

The evening had barely cooled when Dino and I went for a walk.

'I just love my job,' he said. 'Lots of guys stay in but I love to go out. I'll walk anywhere, man. There's nowhere I won't walk.'

We walked under mango trees, through sunlight, under tall palms, passed heat-killed cafés.

'Are you coming to church tomorrow?' he asked. 'Lots of the guys go. Jim goes.'

I said I would not, as I was not Catholic.

'That building sure got shot up,' Dino remarked, in his gentle way. I looked up. The front of the building was a splatter of holes, rips, tears and deep pits, as though it had been hit by a carnival of ordnance.

Dino and Anna had had a lovely day, Dino said, when we met for a drink the next night. Dino was all smiles and Anna was giggly and glowing.

'She showed me all around Brazzaville, and I met her daughter, she's real sweet too. Can you tell her I had a really good day today, and say thank you to her for me again?'

Anna laughed.

'Tell him he is a very nice guy. Tell him I know he will have other girls. I don't care! I like him but if he wants another girl I will go out with a man from Kinshasa. I like him though, he touches me very well. He has a good body. Ha ha ha! Don't tell him that.'

'Don't tell him what?'

'What did she say?'

'She says she likes you but she is worried you will have other girls.'

'Tell her I really like her, she's real beautiful.'

Anna pouted theatrically.

'And tell her I had a real special day today, and I'm really interested in Brazzaville. Tell her I appreciate it. If she could show me things like today I would be really grateful.'

We all went to the club again that night. A woman standing next to me was making notes and chain-smoking fearsome cigarettes. She had a cloud of dark hair and pale skin. Alone of everyone in the night club, she did not seem to care a fig for how she looked, in jeans and a dark sweatshirt. I introduced myself and asked who she was.

'Christine.'

'You are French?'

'Yes. You?'

She narrowed her eyes at the Welsh line, and at the fact I was following swallows.

'Have you seen any?'

'Not in Brazzaville, not yet! What are you writing there?'

'I'm just finishing work.'

'What do you do?'

'I'm a journalist.'

'Oh really? Me too!'

She backed rapidly away from journalism then. She was training journalists, she said, vaguely. I was tipsy and the music very loud.

'Oh I get it!' I shouted. 'You're a spy!'

She laughed, throwing her head back.

'I'm having a real problem not meeting spies in Brazzaville,' I confessed. We drank more whisky, smoked more cigarettes, and then

Christine hit the dance floor. She danced wildly with Anna. Anna and Dino left after a while, but I stayed on. Later, one of the Americans told me a story from another world. I listened in silence.

At the end I said, 'So what do you think happened?'

'I don't know.'

'But he was dead when you touched him?'

'Yeah, I said immediately, this guy is dead. He was limp, there was blood coming out from under his hood.'

'What did the CIA guys say?'

'We don't say CIA, we say OGA – Other Government Agency.'

'What did they say?'

'They didn't say nothing. They just went. Procedure took over then.'

'But they had been screaming questions at him in the shower for half an hour? At a corpse?'

'I guess. I don't even know if he was alive when they dragged him into the bathroom.'

'But you thought he was.'

'Yeah.'

'Do you realise what you have just told me?'

'Yeah.'

'Listen, I've got to ask you. You know, right, that as far as I am concerned, as far as a lot of – people – like me – are concerned, this kind of thing is the enemy. It's everything that is wrong.'

'Yeah.'

'Right. And you believe in America, don't you, you believe that what you are trying to do is for the best?'

'Yeah.'

'OK, so what I want to know is, how bad is it? Are we right to be scared?'

He looked at the floor and his face went dark.

'There're things I've seen I wouldn't even tell a top-cleared American,' he said.

*

Dino had the next day off and, it became apparent after breakfast, planned to spend it in bed with Anna. I went to look for swallows.

'Sometimes there are pythons in the rocks,' said Aimée, my taxi driver and guide. We had driven south of the city, through the Bacongo district, down to the cataracts. We left the car and walked along a track beside the Mambili, a little river that flows through lush greenery to sandbanks, and out into the Congo. The Congo at this point is flowing between two banks which seem much too close together for the immense volume of the river, causing the waters to race and surge up into the gigantic rapids I had seen from the aeroplane: the cataracts.

'That is the car wash,' said Aimée.

Lorries, vans and cars were parked on a beach at the water's edge. Men and boys sluiced them with buckets of water: it reminded me of pictures I had seen of mahouts bathing their elephants. Having parked the taxi in shade we walked down to a wide beach. The sand was deep and soft; pulling your feet out of it was hard going under the Congo sun. In the middle distance were the cataracts, separated from us by the Mambili, a high sandbank and the elbow of rocks between the two rivers in which there might be pythons.

Aimée hailed a ferryman; after a brief negotiation we climbed into his pirogue. I had seen them in the distance from Brazzaville beach: hollowed-out tree trunks up to 30 feet long, gunwales a few inches from the water. The Mambili was deep and fast-flowing at this point: controlling the craft was a methodical, rhythmic miracle of precision. We stuck to the steep-shelving shore, no more than a foot into the flow. The ferryman guided us between rocks and stuck branches, leaving barely an inch between the craft and the obstructions. Standing at the back he paddled with his whole body: a forward bend against the paddle driving us on, minute shifts of weight from one foot to the other moving us fractionally from side to side. I fell into a reverie as we cruised, but then Aimée cried out. He was pointing upwards.

'*Hirondelle!*' he said.

There were three of them, moving fast up the Mambili, heading north-west. They had changed! They were bigger, stronger and moving quickly. Their backs were gun-blue and their tail streamers were longer. Their little red masks stood out, bright in the sun. Their flight was direct. As I watched, one swerved, perhaps taking an insect, but they did not pause in their rush. More came, all heading up the Mambili.

The canoe grounded softly in sand below the python rocks. We began a kind of giant's hopscotch, jumping from rock to rock towards the thundering sound. The rocks rose up to a spine beyond which they curved downstream to form a promontory. Three fishermen were working there, laying out nets, tiny figures, compared to the monsters behind them.

Every second almost a million and a half cubic feet of water gush out of the Congo into the Atlantic, scouring a 100-mile canyon 4,000 feet deep into the ocean bed. Behind the fishermen the other bank seemed about a kilometre away, but now instead of a river the division was a storming sea. Pressure waves 40 feet high erupted in no discernible pattern, their tops a lather of foam. The waters raced like collapsing brown ski-slopes, sliding down in steep falls and then up in explosions of spray. Rearing much higher than the rocks where we stood, the shattered and toppling horizon of the waves seemed sometimes to charge upstream. Between them were broken valleys of air. Through these, through it all, incredibly, the swallows came. They were flying below the level of the spray-bursts, between the cataracts themselves, shooting across the river from the DRC.

I held my breath when I saw the first – surely the bird could not make it through? It did though, slipping easily sideways as the water seemed to plunge around it, climbing slightly and side-stepping the cataracts nearest us. I shook my head and shouted to Aimée who could not hear me above the noise. Then another came, then another; I saw dozens in the half-hour we were there. It was awful to watch to begin with because I was certain one would be caught, but none was; they came towards us, jinked away and followed their forerunners, up the quiet Mambili.

I still do not understand why they did not fly a few feet higher, out of the range of the waters. Perhaps from their perspective the cataracts were legible, their positions possible to fix: the retina of a swallow's eye has two sensitive areas, fovea, which enhance its all-round vision and provide binocular forward vision; birds of prey and humming-birds share this trait, which greatly improves their judgement of distance. But although there were places where the waves were fixed, the gaps between them were not predictable: a smooth patch one second would in the next be an explosive upthrust. Even given that a swallow's understanding of speed, depth of field and time must be vastly different and more subtle than ours is, it still seems an extraordinary choice. There cannot have been many insects to eat amid that chaos: I watched closely through the binoculars and never saw one strike. The only conclusion I came to seems insufficient and certainly unscientific. They seemed to do it merely because they could – as if, moreover, in some wild way, they actually found it fun.

Aimée and I retreated to a coffee bar in Bacongo where we watched building materials being delivered to a Chinese compound.

'They are building everywhere,' said Aimée, with an admiring nod. He was reluctant to talk about himself but keen to point out sights of the city. All he would say of his own circumstances was that he had a wife and children, welcomed the current security situation and needed as much money as I could spare. This evening, we agreed, he would take me to the bus station, Océan du Nord, where I would book a ticket to the north, and the frontier with Cameroon.

Somewhere called the 'Northern Ocean' seemed an appropriate if ominous point of departure for an uncertain journey. When I said I intended to go north overland Philippe, the owner of my hotel and one of the most travelled men I had ever met, shook his head at the timescale.

'There were some bikers who tried to come down that road.'

'How long did it take them?'

'Two months.'

One of Philippe's waiters shook his head too. He was from Sangha province, my destination. It was possible, he said, but not recommended. It could take a long time. There had been more than sufficient rain to close the road.

'You must extend your visa,' he said.

The office granting extensions was guarded by hustlers. When I finally broke through them it emerged that the minimum extension was three months, and the price was prohibitive. Mulling this over, I went out for lunch with Christine, the Frenchwoman who had laughed when I accused her of being a spy.

From 1940 to 1943 Brazzaville was the capital of Free France. De Gaulle gave a speech here in 1944, in which he said moral and material progress was dependent on the fortunes of people 'living on the earth where they were born' and the stake they held in managing their own affairs. This made him popular in Congo. And it was easy to see why Congo was popular with the French. Sitting under the pavement cloister of another Lebanese restaurant with Christine, watching her order from the waiter with truly Parisian specification, then lunching on steak frites with green beans, a glass of red and a little coffee to follow, you could understand why de Gaulle and France hung on to Brazzaville for another decade and a half after that speech, refusing to relinquish it until 1960, the year independence movements swept half of Africa and put Africans, nominally at least, in charge of their own lands.

'Did you see Sarkozy's speech?' I asked her. The president, notwithstanding rafts of headlines in the Paris press along the lines of 'Sarko's year of madness' (it was claimed he promised much and delivered little) had made a speech to the South African parliament in which he promised a new transparency in France's dealings with Africa.

Christine blew cigarette smoke. I asked about the current state of Congo.

'They do not have water, and yet that is the Congo River down there! There is one factory that still works in the whole country. One!'

'So what is going to change?'

'Well, have you seen the Chinese?'

'Yes.'

'Do you know what is the difference between the Chinese and all the other countries who have people here?'

'What?'

'Everyone else drives Land Cruisers. The Chinese walk.'

The implication was that the Chinese were unafraid, or felt no need to distance themselves from the population, that they were tough and streetwise to the Congo. The difference between those who walk and those who drive seemed one way of distinguishing between those who were there to stay and those who were passing through.

On my last night in Brazzaville Dino and I went to our favourite bar again, where we sat at the counter drinking beer. I had just refused to buy a bottle for a young woman when there was a tap on my left shoulder. The raucous old lady who sat at the next bar stool grinned at me, her face wrinkling into dozens of smiling lines. Her hair was either cut to nothing or she was bald. Her eyes were bright sparks in the dim of the bar.

'May I have a beer?' she said.

'Of course, Madame,' I answered, without thinking.

She did not exactly fall off her stool, but she was pleasantly surprised. The bottle appeared, the top was whipped off and the old lady sank a deep draught. She set the bottle down, put her hand back on my shoulder.

'I give you the protection of God,' she said. 'I give you the protection of God for all your travels in Africa.'

I thought of her later that night, in the upmarket hotel room (Philippe's being full), with all my stuff emptied out of the rucksack and the table covered in maps. I decided she was my witch.

The ticket was booked: Océan du Nord was a rough mud compound on the edge of town, but the buses had been reassuringly formidable; they had tyres like giant Land Rovers. As far as I could establish, the route involved a day-long bus ride, a night stop, a taxi of some sort, a walk which was supposed to be 30 kilometres through the forest, a canoe and another taxi to the frontier at Ouesso. A motor pirogue up another river would complete the journey to Cameroon. What happened then was not clear.

The problem was the swallows. On the bearing the birds had been flying, their route would take them well to the west of mine, across the Pool region and Nibolek province into Gabon.

Enquiries about Gabon were met with derisive laughter. The Gabonese had taken the French too seriously, the Congolese said; bureaucracy was a religion with them.

'If you have not got a visa and a letter of invitation forget it,' my sources said. The Africa guide concurred. I thought about it for a long time. Head for Gabon and hope, or curve up-country, and trust?

An aeroplane took off from the airport. I went to the window to watch it go. The plane was huge, Russian, I thought. I wondered what they were carrying. Coltan, the diamonds of the mobile phone age? Technicians, advisers, spies?

'We hate the Russians,' Christelle had said, on behalf of her friends. Christelle was in her mid-twenties, from Ivory Coast. She seemed to live for music: when there was none playing, in a pause, Christelle would dance anyway, to tunes only she could hear. She partied at night and slept in the day, rising in the evening to eat. We had danced together, drunk together and been lovers. We had met on the dance floor of the night club. Ivory Coast had nothing for her, Christelle said. She liked Brazzaville because it was peaceful. She said she planned to go to college – or get training – or work – at some point. She was vague about her future but apparently unworried by it. I do not know if our liaison was any different from a million holiday romances which take place all over the world, all the time: it was impossible to tell whether I was desirable to her because I was white and rich or simply for myself, but in the three days we spent together Christelle asked for nothing. I think she would have been surprised if I had asked her to pay for her own dinner, but I would not have done, anyway. The innocence of our relationship, notwithstanding its limited lifespan and therefore casual nature, had not made us any less self-conscious, when we had been out to cafés and restaurants. I imagined that women looked at Christelle and men at me with the same assumptions. She must be in it for the money, me for the sex. Christelle assumed a dignified hauteur, when heads turned, and rolled her eyes at me.

Buying condoms in Poto Poto was hilarious: the chemist presented a packet and tried not to laugh. I deciphered it.

'Are these – luminous?'

'Yes!'

'Thank you – do you have – anything – normal? I hope I will be able to find it without glowing in the dark . . .'

'What's wrong with the Russians?' I had asked Christelle.

It turned out that there was one among the aircrew of the regular Brazzaville–Moscow run who was 'a pig'. One night he and his friends had taken some of Christelle's friends back to a hotel room, got drunk, the pig had lost his temper and beaten a girl up. Brazzaville would not forget.

Christelle was asleep in my bed now; she had turned up, eaten room-service sandwiches and crashed out. She did not stir in her sleep. In Europe we would count as a four-night stand, spread over a week; here, as far as the hotel staff were concerned, she was a prostitute and I, if not a pig, was otherwise in the same bracket as the infamous Russian. Outside there was a roll of thunder. Lightning flashed and the thunder came again, louder now.

'Kinshasa always sends us their rain,' Aimée had said. I enquired about going there too, but Anna had forbidden me.

'You cannot go there because I cannot take time off to take you and Dino.'

'Why can't I go anyway? There are lots of westerners there, aren't there?'

'Yes but they are all with organisations. On your own you will be eaten alive. Kinshasa is mad, do you understand? We come to Brazzaville for a rest!'

Christelle slept soundly as the storm came on. Soon there was a hissing, then wild smashing torrents of rain. Philippe had said he spent most of his time repairing and repainting: now I could see why. The palm trees bent over and whipped furiously in the wind as water battered the window in jets like a pressure hose. It was exhilarating and ferocious and I knew I would not sleep; there were only a couple of hours left before Aimée came to take me to the bus. I wondered what the rain was doing to the roads and considered the promised walk. It was not clear how long it was – some said 30 kilometres, some said 50.

I woke Christelle at quarter to five, briefly, to say goodbye. Half of my rucksack had been transferred to the chair: shirts, a lightweight jumper, a pair of shorts, maps of the Congo and Zambia, books I had acquired but could not carry, socks, boxer shorts, T-shirts. I asked her if she could use it all and Christelle nodded. She took my hand and squeezed it, looking hard into my eyes. I promised to tell reception to let her sleep and she nodded. I told her, in a pathetic way, to take care, and she nodded again.

The sky greyed as Aimée drove me through the rush hour, which started with the light. Océan du Nord was a scrum of people and bags. Aimée bade me good luck and drove away. I sought the Ouesso bus. Beside the two luxurious all-terrain charabancs which I had seen the day before was a beaten-up, lopsided Toyota minibus. The charabancs were going only as far as Oyo, on the notoriously good roads which led to the president's town. The little scrap-heap was going as far as Makoua where the Ouesso road ran out.

The sun may have risen behind the clouds but it was as dull and hot and close as fever. I stood swaying, dozing on my feet, retaining consciousness principally through curiosity as to how all the travellers and our bags could possibly be crammed in. It took an hour. In the end the front row of seats were full and there was an amazing overhang of cargo suspended above the second row, including two satellite dishes, sacks of rice and heavy-duty electric cable. We would be travelling 'cinq par cinq': five abreast in five rows of seats. I was placed by a window. My neighbour was a patient lady with a comfortably well-covered flank. We were each issued with a roll and a carton of juice, and then we were off.

It was not a pretty journey. The land north of Brazzaville had been cleared to nothing, to yellowish grass, dull under heavy skies. Among ridges and shallow valleys villages put up smears of smoke. Further north again there were clumps of trees, and many hundreds of stumps. I fell in and out of sleep and woke to a clamour: passengers howling at the driver who could not hear them through the densely packed luggage. Eventually a message reached him and we stopped. The aspect of the day had changed: now there were higher trees, taller grasses, and the sun had come out. I extracted myself through the window, smoked half a cigarette and looked for swallows. There were none.

Travelling five by five was an endurance test, I learned. With a skinny European ass your only defence was to keep shifting your weight from buttock to buttock; with something more substantial, I brooded, enviously, you would be cushioned like my neighbour. More significant movements, leaning forward or back, affected her and therefore the person squeezed onto her other side, so these you kept to a minimum. Slight shifts of a leg would make fifteen minutes' worth of difference to the numbness of your side or the pain in your lower back. Sleep, if you could catch it, was a saviour.

We came to Makoua at the most lovely time of the evening, about an hour before sunset. The bus pulled up under a tree. We had been skidding and rattling over sand for the last hour through ever more wooded plains. We hauled ourselves out and stood on soft sand as the bags were freed from their restraints. Makoua appeared to be a collection of small bars made of wood and corrugated iron, a little roundabout, a road running on ahead and not much else at all. There was a peace in the air like a protracted siesta. Children looked at us and a man in a bar raised a bottle.

'Is there a hotel or anything?' I asked.

'Yes!'

A smiling young man, darker-skinned than anyone else, came forward.

'There is a place . . .'

We fell into conversation.

'I am PJ,' he said. 'Everyone in town knows me.'

'Are you from Makoua then?'

'No, I am from Cameroon.'

'Cameroon! What are you doing here?'

'I am travelling to Gabon and Sao Tome.'

'Really? Why are you going there?'

'Because I know it is beautiful. There are beautiful people in Sao Tome.'

Sao Tome and Principe are two islands off the coast of Gabon. They were discovered by the Portuguese in the fifteenth century. The guidebook showed a Caribbean-type island. I do not know what inspired PJ's enthusiasm for the place, one of the smallest countries in the world, but as we talked it became clear that for him it was a promised land, a dreamscape.

We turned off the road, up a track, and passed a football pitch where dozens of children were engrossed. The hotel was half-finished, clean and comfortable. I was issued with a bucket and a hurricane lamp. On my way back to the centre to meet PJ for dinner another man stopped me.

'I am a teacher,' he said, 'but there is no job for me.'

'How do you live?'

'Ah,' he said, with a grimace, 'I am working on the roads.'

There was no electricity in Makoua except from individual generators. PJ and I sat in a hot darkness; he was invisible but for the whites of his eyes.

'I journeyed down here through the forest,' he said. 'At the border they took my passport and all of my money. I was a Rasta then.'

'What happened to your dreds?'

'I cut them off because of the hassle! But I am still the only Rasta in

Makoua. People are very kind to me. I have many friends here because I am a friend to everyone.'

'How will you get to Gabon and Sao Tome?'

'I am working, building a school. I save a little money.'

Inside a shack, by the light of a hurricane lamp we ate bread and scrambled eggs. The chef was from Mauritania and limped from a recent injury involving boiling coffee. I promised to return the next day with burn cream. The stars came out and fireflies winked above our table.

'Ha!' said PJ, when I exclaimed at them. 'In Cameroon they call people from the north fire-flies, because we are darker than everyone else.'

The night of Makoua was utterly unfamiliar, a blackness which gathered around, pressing in on the oil lamps and squeezing the fireflies so that their flashings were like the blinks of buoys far out to sea. The fatigue of the road, the peace of Makoua and the smiling chatter of PJ made me languid with relaxation. I would like to live here, I thought. I could imagine setting myself up in a low house on the edge of the forest, with hurricane lamps and books, living a slow life.

PJ talked about Sao Tome, and we discussed travel and our lives. All we wanted, we agreed, was to live somewhere beautiful, with someone to love, and friends to visit.

We met for breakfast in the morning, cooked by the Mauritanian chef, who accepted burn cream and antiseptic for his leg. PJ introduced me to Judicael, a young man who was going to be a radio journalist when the radio station was working. He and PJ were great friends: they did the full handshake, shoulder-bump and salute, then taught it to me. By the time it was completed we were all laughing like old friends. At the moment the radio station was a single computer and a large stereo but Judicael said he hoped it would not be long before they were able to start work. For now he was playing around with the computer: fifteen years after email first appeared in Britain it was strange to witness its arrival in Makoua, as he and PJ signed up and obtained their first addresses.

We walked slowly down to the river and the bridge. The water was a silted, sludgy green, the bridge was an substantial span in iron and concrete and the air all around it was full of swifts, and no swallows. That was my road, PJ said, the road north, the way he had come.

'You will see Pygmies!' he said. 'You will see the forest. You will have a truly wonderful journey.'

Standing guard in front of the bridge was a sign in English and French.

DANGER — EBOLA

DO NOT TOUCH ANY

DEAD ANIMAL IN

THIS FOREST

I studied the sign with a kind of horrified thrill. Even by the standards of African fates, Ebola, named after the Congolese river where it was first identified, is a horrible way to go. Vomiting, difficulty breathing, pain, bloody diarrhoea and bleeding from the nose, mouth and anus is followed in ninety per cent of cases by death. There is no vaccine or cure. No dead animals then, I told myself. I will have nothing to do with them, whatever happens.

It was already hot and we made our way back more slowly. PJ took me via the school he was building, a skeleton with a partial roof, standing in red mud and yellow sand.

'That is where I work,' he said.

'When is it going to be finished?'

'We do not know. The materials have not arrived.'

'So what do you do?'

'We wait. But I am proud that one day children will go to school there.'

PJ told me his story in several different ways, in a different order each time, rearranging the facts so that the emphasis fell equally heavily on each of them.

'I came down from Cameroon because I want to go to Sao Tome. I hope if it is God's will to find someone to marry there, to live there, because it is beautiful and the people are very good – they are a great mixture of different kinds of people. But I have lost everything – everything. The police at the border took my money, they took my passport, and so I am trapped in Makoua. I do not even have shoes.'

PJ hailed a tall, spare man who left a crowd of milling workmen, and introduced us.

'Jean is from Senegal!' PJ said.

'What brings you here?' I asked.

'I came down from Senegal looking for work,' Jean said. 'I am an English teacher. But there is no work for teachers – there is no school. At the border they took everything from me – my money, my passport, everything. All I want to do now is get back to Senegal.'

'How do you survive?'

'I work on the roads,' Jean said. He looked at the hot ground, his expression dim. Then he raised his chin to me.

'Beware the Ides of March,' he said, and did not smile.

On the way back to the middle of the village we walked along a stretch where one of the labour gangs was working. There were two or three large machines, bulldozers and lorries and perhaps sixty men spaced out, a couple of metres between them, in a ditch of earth and broken stones at the side of the road. With picks and shovels they were deepening and carving this ditch into a rectangular channel which was then being concreted. The labour of their task, under the ten o'clock sun, was all too easy to see. The earth and stones did not break easily. Each shovel-load cost a grunt and a heave. The men sweated and did not pause. Fifty yards away there were two Chinese overseers in cotton shirts and sun hats. One of the workers looked up and cried out: the teacher I met last night. His face broke into a wide smile through the sweat and he leaned briefly on his shovel as we shook hands.

'How are you?' he cried.

'I am very well – how are you?'

'Hot!' he said.

As we talked I looked down the road and saw all the labourers for what they might well have been. Teachers, students, journalists, travellers, writers, artists, thinkers, doctors, accountants, business-men – all without papers, all without money, from a dozen different countries and all with picks and shovels under the sun, working with who-knew-what hope of change: the Congo's new slave labour.

PJ accompanied me back to the hotel, where I packed. His eyes fastened on my unused shoes. They were lightweight and waterproof with wonderful treads, like a network of little suckers. They had ridden in my rucksack in anticipation of today: I had bought them in Cape Town with the intention of blooding them in Congo.

'They are beautiful, your shoes,' PJ said.

I was embarrassed by my freedom, by the wealth of my equipment, by the stamps in my passport, the cash in my wallet and the power in my plastic cards. PJ had not expressed envy for any of these things but in a minute he would walk me to where the taxi was loading and watch me go. Clearly I must give him something for his kindness and hospitality.

'You have them,' I said.

'You are sure?'

'Yes!' I said, convincing myself.

PJ put them on and we walked back to the roundabout. In front of a little bar was parked a Land Cruiser in an advanced state of dilapidation. Around it were faces I recognised from yesterday's minibus: a young husband and wife and their small boy; a laughing woman with several sacks; an old lady with a tubular basket; a gentle-man, who sweated under more weight than the rest of us; and a young man. We all said hello and I paid for a seat. As a white, therefore rich, it was assumed I would travel in the cab with the driver, for a small supplementary fee. The sweating gentleman did not look happy: there would now be a crush in the cab. The driver was a big rangy man in very tattered clothing. He looked as battered as his vehicle.

When I first saw it, I assumed nothing but the worst treatment and the poorest maintenance could have caused the Land Cruiser such damage. The driver was not happy with it either: there was a jagged diagonal scar across the eye of the left rear wheel where a bolt should have been. The driver had a bolt of approximately the right size: he held it to the scar, demonstrating to his motorboy how it might fit, holding it to the wound in a rather unlikely way. The motorboy nodded gravely. Every truck and taxi has a motorboy, a driver's apprentice. Some drivers, commensurate with their status, have two.

An hour after the announced departure time, we set off. There is a strange temporal alchemy of central African travel, which means that everything leaves at least an hour late, but nevertheless always arrives on time.

PJ waved goodbye and wished me Godspeed for the road. I wished him the same. '*Mettez vos chausseurs sur les bonnes routes!*' I cried. Put your shoes on good roads. We held up our hands in a long salute.

The Land Cruiser ground and bounced out of town, taking two fairly rough diversions to avoid the roads under construction by the gangs. I was squeezed between the driver and the gentleman, who revealed that he worked for the council in Ouesso. When I asked him what that entailed, 'Development' was all he would say.

We passed the Ebola warning and crossed the bridge. The gentleman nodded at the thickening trees.

'Today you will see true equatorial rainforest,' he said.

No more than twenty minutes out of town the rainforest began. It was not unbroken: here and there were clearings of long grass. There were people on the road pushing bicycles, dwarfed, away from the clearings, by the immense height of the trees. The forest formed a thick and battered wall on either side of the track. Its fringes were ravaged by tyre-tracks and piles of logs, slashed bushes and smeared heaps of ash. Very quickly the road became a track, which in turn became something else: I began to understand. It was not its handling nor its maintenance that had almost destroyed the Land Cruiser: the driver was wonderfully adept, sensitive, skilful and quick. The problem was the track, which was a killer.

A rare stretch of simple hard sand ran into a wall of trees. A single machine, a dragon-like thing with a long clawed arm, wrestled with a tree. Beyond the tree was a village, and beyond the village, nothing but forest.

We stopped. A man appeared, then another, dragging a generator. A third man emerged from a hut carrying a car battery, another brought a welder and goggles. In not much more time than it took all of us to get out of the Land Cruiser the battery was connected to the generator, the welder plugged in, the goggles put on, the bolt welded to the scar. The mechanical dragon won its battle with the tree, which came down with a rustling crash. It was midday exactly and a chicken, standing still, apparently watching proceedings, cast only the merest shadow directly beneath itself. We all climbed back into the Land Cruiser, crossed the equator and plunged into the forest.

Crushed between the hot gentleman and the driver's elbow I tried to be as small as possible. The Land Cruiser's gearbox had lost whatever had once covered it and now gave off blasts of heat, steam-cleaning my left knee. We stopped occasionally to pour water into the engine. The driver coaxed us up steep banks of mud and plunged us down into ravines. In some places the track disappeared entirely into long lakes of standing brown water, their sides knitted tight with trees: the Land Cruiser forged into the lakes, becoming a kind of barge. It was an indomitable machine, the toiling engine putting up fresh hot blasts of vapour like whale-spouts when we pushed through water. We passed streams and dark green pools. As the machine swayed from side to side, roared, groaned and fought its way forward I tried not to think about what would happen when, as seemed likely, the gearbox exploded. On downhill stretches the driver accelerated to 50, then 60 km/h, all of us hanging on grimly as the wheels bucked over stream beds. Where we hit landslips and impassable craters the vehicle heeled over to a terrifying degree and proceeded as if on two wheels.

It was an epic, operatic performance which ended with a gentle run up into a village. It was mid-afternoon and very quiet. The village was a scattering of wood-framed huts, roofed with a kind of thatch. The Land Cruiser stopped.

'C'est ça,' said the driver. That's it.

We climbed out. Under a tree stood a group of unsmiling young men with bicycles. For a price they would pedal your bag through the forest to the river. There was something thuggish about them, as they bartered with the travellers. The young man travelling alone drew me aside.

'I am Bertrand,' he said. 'Who are you?'

We shook hands.

'What happens now?'

'Now we walk!'

'Right!'

'But first,' he said, 'come with me.'

'Where are we going?'

'Just behind that hut, into the forest.'

'Why?'

'To smoke some *tabac congolais*, to give us force.'

Bertrand produced a small knot of green weed and an empty cigarette packet, which he tore open to make a flat rectangle. Then he worried one edge of the card and blew on it with short fierce puffs. One of the layers of paper of which it was composed began to peel away. Carefully, with more worrying and more blowing, Bertrand stripped it off.

'*Et voilà!*' he said, and rolled it into a joint. When we emerged from behind the hut we were both charged up and giggly. The bicycle men and the other travellers had disappeared. We walked into the trees.

The rainforest grew up in canopies around us, a pavilion of widening, ever-higher green tents. You could not see very far into it; as the plants pushed up they led your gaze up with them, to the sky. It was like being a shrimp at the bottom of a deep green moat. At first we went quite slowly as I marvelled and Bertrand laughed at my amazement.

'Look!' I cried, '*Regarde-moi ça! Le papillon!*'

The butterfly was as big as the palm of my hand and deep silken crimson.

'But I have never seen that colour before!' I breathed. The insect swam across the path in front of us.

'You wait until we meet the Pygmies,' Bertrand said. 'You haven't seen them before, either.'

We came into clouds of butterflies; lemon yellow, orange, blue and white, lifting off the path like shoals of blossom. The track narrowed to a single path in places and the forest was very quiet, there was barely a bird call and no wind. I noticed twigs bent into strange configurations, crosses and triangles, and wondered if they were Pygmy signs. A person standing still 5 metres away would be invisible to us and though we did not say anything we both felt the forest watching. We came to a stream where small children were playing: they pointed at me excitedly and Bertrand said something which made them laugh.

'What was that?'

'I said the White was going to turn pink!'

'Oh, ha ha!'

The track climbed out of the trees and took a long curve across a stretch of open ground. The temperature leapt, away from the shade, and the heat was a solid weight. We took turns singing to keep our spirits up. Bertrand hummed and la-la-ed to himself. I sang a random medley ('Bread of Heaven', 'The Big Ship Sails on the Alley-Alley-Oh', 'A Fox Went out on a Winter's Night', 'Diamonds on the Soles of Her Shoes'), then Bertrand talked a spoken equivalent, a free-form chatter of whatever came into his head.

'My father is in the police at Ouesso. You are going to love it there! There will be women and lots to drink. We are going to have some fun! Oh, you White. You have no idea. Do you know what was here a few months ago? Ninjas! A few years ago it was all war here – you would not have been able to walk along this path singing your songs – *mais non!* They would kill you quick. Do you know what it is like in Congo? No, you do not know. Have you heard of the Colonel? I don't think so. Listen, there is a colonel. He is in charge of security. If the Colonel asks you a question you answer fast. Because if you do not answer fast the Colonel takes out his pistol. And if the Colonel takes out his pistol he does not put it away until he has used it.'

'What's his name?'

'Don't worry about that.'

'OK. Aren't you hot?'

'Yes, but you are hotter. Let's have some more water.'

'Haven't you got a hat?'

'No.'

'Wait . . .'

We proceeded, Bertrand wearing one of my shirts wrapped round his head like a turban. We passed some women coming the other way. One of them said something which made Bertrand crow with laughter.

'What was that?'

'She said the White is turning pink!'

Now the track descended again and the forest reared up ahead. There were huge trees, their great red trunks towering as straight as telephone poles, as wide as oaks at the base.

The water we drank turned straight to sweat. The path now returned us into the forest; we seemed to tunnel into its dark shade. Now we began to cross wide dark streams, bridged by U-section iron rails. Bicycling across one would be a severe test of nerve; Bertrand and I competed, covertly, in our demonstrations of nonchalance. We came to a village. A group of women were sitting on the ground.

'See?'

'What?'

'Pygmies,' Bertrand whispered.

'*Bonjour!*' I said, and raised my hat.

The women nodded and smiled. They were small, rather than tiny. Clumps of pineapples grew like brambles beside the path and here and there were clusters of bright flowers. A man came slowly up the path towards us.

'*C'est un Pygmée,*' Bertrand whispered, unnecessarily.

'*Bonjour!*'

The man nodded gravely.

We caught up with two of our fellow travellers, the young wife and the old lady, who had stopped to rest. As well as her basket the old lady was carrying a satchel: after some protest she let me carry this for her – she did not speak French and Bertrand grinned at our exchange rather than interpreting her Lingala. Her feet were bare, her basket loaded, but her pace was just as fast as mine. Insects swam in lazy clouds at the stream-crossings and I mourned my donated shoes. A blister was beginning to grow on the ball of my right foot. I tried going barefoot but the grains of sand on the hot path prickled under my feet like glass dust. The old lady took up station behind me and gently shooed me along. Bertrand and the young wife chatted in Lingala and we sucked our teeth and laughed a little nervously as the balance-beam water crossings became harder. The young wife giggled and skipped across them as surely as if she were wearing a safety harness. Another DANGER EBOLA sign appeared, my blister burst, water filled one shoe and I wondered in how many ways the disease could be transmitted.

No one seemed to know how far we had to walk.

'About 15 more,' Bertrand said.

'But you have done it before?'

'Yes. Another hour or two.'

We emerged again into an area of high grasses, the trees falling back briefly. There was a flicker of movement above us and I looked up.

'Bertrand!' I cried. A single swallow shot over our heads and pulled round in a long, controlled skid.

It circled us once and swept away, slightly to the west of our path. The hair seemed to stand up on my arms and I felt flushed from my stomach to the back of my head with a strange, tingling warmth, as though the bird had cast a spell around us. It gave a brief, chittering call as it turned its circle over our heads.

'Did you see that? An *hirondelle* – that's the bird I am following! And it was here – just here!'

Bertrand grinned, half in encouragement, half in mystification. I was suddenly swept with a wonderful kind of euphoria, an electrifying feeling beside which the mild buzz of the dope, earlier, was as nothing. The whole day, this journey, the decision to come this way, all were cast suddenly in a different light; a different substance in my mind; as if a tentative hope had suddenly hardened into certainty. Suddenly it did not matter that I was so far away from the world as I had known it. This remoteness, the commitment to this track, the helplessness, in which I could only follow, only keep going – all were justified, were confirmed. It was as if I had been blessed.

We came down the Likouala River as the sun was beginning to lose its grip on the sky. It was still crushingly hot and humid, but the lapping of the deep grey-brown water seemed to radiate relief. The bicycle men were sitting in the shade with the luggage: one asked for a supplementary tax from Bertrand, and there was a debate, which he lost.

'Now I have nothing,' he said, sadly, but still smiling.

Having carried my own bag, I was still flush with CFA francs.

'You bargain, I'll pay,' I muttered. There were two pirogues waiting on the bank. We loaded slowly, carefully, everyone sweating and heavy-legged. I swept my hat through the water and crammed it onto my head.

'The White is hot!' laughed our paddler, and pushed us out into the stream. The river unravelled slowly around a bend. There were no villages, no paths, nothing but the Likouala gliding through the immense press of trees. We climbed out on the far bank and followed a sandy path to a village. There, sitting under a tree, pointing away from us, towards the north, was another Land Cruiser. This one was in excellent condition, with a heavy-duty cage covering its open back. In a hut, in near-darkness, sitting at a table with a cold bottle of beer open in front of him and dark mirror-shades covering his eyes, was its owner and driver, on whose pleasure we all now depended. This was Chef.

We went in two at a time. When Bertrand and I entered, Chef bade us sit with a nod of his head. He can have seen very little from behind his shades: to the naked eye there was almost nothing visible in the gloom but eight bright bars of sunlight glowing through slats. Chef required that we drink with him (a sort of alcoholic grape juice from a carton, for us, more beer for him), converse in a civilised way and accept his price. Once that was achieved, he nodded us away: Chef did not handle the money; that was his motorboy's department.

Outside I fell into conversation with a small man in a bright red baseball cap.

'So do you like our forest?'

'Yes! It's incredible!'

'You are a tourist?'

'Sort of . . .'

'Ah, swallows,' he said, when I had explained. He had no particular stories about them. Like many Africans I met, he could name them, and say when they came and went, and that was all.

'So you are not here to see us?' he smiled.

'You?'

'Pygmies!'

'Well, it is very good to meet you. Do many people come to see you?'

'I am a guide to the reserve.'

'Oh yes?'

'Yes, we are guides for French tourists, and other kinds too.'

'Is it very expensive to get in?'

'Ah! But you are in!'

He explained that tourists were flown in and out, staying in a luxury hotel. I asked him why a reserve was necessary.

'Because they are killing the forest,' he said. He pointed out and named the different kinds of giants whose heads we could see from where we stood. Iroko, known as the African teak, and sapelli, the most magnificent of all, formed an entirely separate realm; green pavilions on red trunks, watchtowers above the forest, a canopy above the canopy.

'To us they are sacred,' he said, 'especially sapelli.'

The countries of the north and west are greedy for their corpses. Parquet floors, guitars, even the interiors of Cadillacs require them. I had not understood that the ancient forest we stood in was one of the few significant tracts remaining in the Republic of Congo: to the west and south of us millions of hectares had long since been destroyed.

'There is still forest in Gabon, and some in Cameroon,' he said. 'But you will see.'

Two motorboys loaded us all into the Land Cruiser. I opted for the cage.

'It's four hours,' said the motorboy. 'You're sure?'

'Yes he's sure!' cried Bertrand, who had topped up with *tabac congolais*. 'Chef is just having some *tabac* too,' Bertrand said, twinkling, 'to give him force.'

'Well, better stoned than drunk,' I said uneasily, looking at the Land Cruiser and judging it capable of lethal speed.

'Chef is the best driver,' Bertrand said, seriously, having met the man only twenty minutes before.

'We will confirm that when we arrive,' I replied.

'Is the White scared?' asked the laughing lady, whose name was Mariam. She had a great sack of fruit of which she was even more protective than was the young wife of her son. Before I could think of an adequate riposte, knowing that yes or no would provoke a gust of Mariam's particularly powerful laugh, the motorboy appeared and swung the charred spine, ribs and belly of a pig onto my feet, followed by two sacks of some sort of bean. He climbed into the cab and drove us to the other end of the village where, exuding immense authority, Chef appeared, shooed the motorboy out of his seat, and we set off, with blasts of our horn and salutes to the waving Pygmies.

We raced down a red earth track at about 90 km/h. Bertrand whooped and banged down on the roof of the cab. I grimaced, at which he laughed and banged again. Three or four miles outside the village we stopped. Chef dismounted and came around to the back. We all fell silent as he fixed his gaze on me.

'So, are you going to give me a present?' he enquired.

'A present?'

'Yes. Are you going to give me your hat?'

There were titters from the travellers. This was theatre, I realised: the point was to entertain.

'Right,' I said, casually. 'You're the driver of this thing, are you?'

'*Eh, OUI –* !' said Chef, to the gallery: this was the most stupid question on earth.

'Good. *Alors*, if you drive well, incredibly well, perfectly, in fact, and if you get us to Ouesso very safely and very comfortably, and if we have enjoyed our journey and never felt that we might die, and if I am in a good mood when we get there – a truly superb mood – then it is just possible that I might give you a present but I can absolutely promise you it WILL NOT BE MY HAT! This is MY HAT, CHEF!'

The travellers cheered. Laughing, in a concessionary way, Chef climbed back into the cab. It was satisfying but also vaguely unsettling; clearly, and somehow on behalf of the travellers, I had won round one; equally clearly there were more rounds to come. We raced on. Bird-watching Congolese style meant standing at the back of the cage alongside the young husband: the two motorboys and Bertrand took the front, near the cab, leaving what space there was among the cargo in the well of the cage free for the women and children. We ducked and swayed as branches and chicottes whipped towards us at 100 km/h: the chicottes are long feathered grasses like shoots of bamboo.

'Swift!' I announced, or 'Swallow!', though I did not see many of them.

'Toucan!' shouted the young husband, pointing at a hornbill.

'*Chauffe! Chauffe!* Chef!' howled Bertrand, hammering on the roof of the cab – Hot! Hot! – as he urged ever greater speeds, grinning hugely at my discomfort.

'I'm bloody serious!' I shouted at him. 'I want to see these birds!'

The travellers burst into loud comment and I blushed with realisation – for the first time in days I had spoken in English. They laughed. It was as though in speaking my own language by mistake I had fully revealed myself, and they approved. We stopped again. The motorboys jumped down and Chef appeared, hands on hips, directing them to a particular bush, from which they withdrew the body of a fat black pig with maggots oozing out of its head. This they dragged to the back of the cage. The cooked portions of the other animal were shifted and the corpse was heaved up over the tailgate and onto my feet. We all edged away from it as much as possible. It was difficult not to imagine Ebola, like an evil smell, penetrating my shoe, sniffing around my blister and worming its way into the flesh.

Chef stopped again, another 10 kilometres on, in a village on a hillside. The light was going now; the last swifts and swallows hawked overhead and then vanished as first stars, then fireflies, appeared. Somewhere in the village, Chef was in protracted negotiation, eating, and no doubt drinking and smoking. Bertrand and the motorboys gossiped in Lingala, the young wife fed her boy and Mariam passed around bananas. When it was quite dark, the night thick and pressing on the forest floor, paling only among the stars, Chef reappeared, still wearing his shades and now carrying a powerful torch. With him were a man with a machete and a very small boy. At a command from Chef the motorboys seized the dead pig and laid it down on the track. The man with the machete approached it with great caution.

'Ebola?' he said.

'No, this is not Ebola,' Chef asserted. He stretched the animal's legs and tilted them this way and that, bathing its armpits and stomach in the light of the torch. As his father talked doubtfully with Chef the little boy rushed forward, gurgling with delight, grabbed the pig's front trotters and braced his feet against its shoulders. For a moment we all fell silent and stared. The little boy babbled, urging his father on. I believe we all had the same thought: if the pig had Ebola then the beautiful little boy now surely had it too. If he has it, I thought, I do not care if I get it. I do not care about anything any more, if that lovely little boy has been infected; we might as well all die together.

The father relented and bent forward. He stroked the blade of the machete down the centre of the pig's chest and the skin split under it. The butchery was completed with wonderful speed and skill and the pig was soon returned to the cage, now in hams, securely tied in sacks. A sack of mangoes was heaved up to join us and we set off again, Chef's mobile trading post hurtling like a single bright spark between the toes of the forest.

We huddled down as it grew colder, into whatever space we could find. We took turns singing, as the floor of the cage bashed up and down, then we fell silent. A long time passed slowly, then there was a commotion and something was thrust at me. The old woman held something flapping in her hands. I switched on my torch. Brown- and cream-barred feathers, long wings and the stunned dark eyes: it was a nightjar, a beautiful bird, the first I had ever seen. I was reaching out for it when the old woman lost patience and flung it over her shoulder, out of the cage. Bird-watching Congolese style . . .

The forest did not fall back as we descended towards Ouesso, but the darkness did. Villages which had been dark now had oil lights. Then a generator, and electricity, then the blue firelight-flicker of television. Low dark buildings came up around us and side roads, other vehicles and motor scooters, and at last we stopped beside a mass of people milling around a bar. Euphoric to have arrived at last I jumped down, into the hands of the police.

They demanded my passport with a drunken belligerence. Halfway through an argument about the cost of not having a non-existent tourist permit Chef drove away, taking all the travellers except Bertrand, and, I realised too late, my notebook. In a foul mood and many thousand CFAs poorer I placed myself with very bad grace in the hands of a man who said Bertrand and I were lucky to have found him because he ran Ouesso's only good hotel. Bertrand confessed that his father did not actually live in Ouesso, exactly, but rather a little way downriver, and wondered if there might be sufficient CFA to get him a bed in a hostel.

The rain came again that night, with a lightning storm. In the morning Ouesso resembled a film set of small buildings sinking into fields of reddish mud. Car wheels spun and skidded and people picked their ways between mires with the concentration of hopscotch players. The kind hotelier showed me a stationery shop where the proprietor changed euros into CFAs, and a man in a splattered red car pulled up alongside us, and grinned. We shook hands and exchanged pleasantries. He drove off.

'Who was that?'

'That was the man who owns the hotel,' the hotelier said. 'He is a very important person in Ouesso; he is in the customs. But now he knows who you are, you will be fine.'

'He has made a lot of money in customs?'

'Yes, a lot of money.'

There was a very strange atmosphere in the town. People went to work, or out into the day, at any rate, in a rather unconvinced way, as if this was what normality demanded of them: the unconvincing thing was the normality itself. As far as I could tell, the only way of life in a place like Ouesso was to attach yourself to something or someone with access to money and then wait, hoping some portion of it would trickle to you. Apart from the few shops and bars there was no obvious income available to the townspeople but service in the police or the army: a great many men were in uniform. Everybody else seemed to congregate around the port.

Ouesso stands on the banks of the Sangha River, which flows eventually into the Congo. Ouesso's port principally serves motorised pirogues, canoes with outboard engines, but the Sangha here is a busy waterway and the riverside was crowded with people. Among them were the old woman, the young husband, wife and child, and Mariam, holding my notebook aloft.

It was a joyful reunion. Mariam had found my emergency €50 note, concealed between the notebook's binding and cover, but she could not read the English notice I had printed in the front, promising €50 for its safe return. While I went to deal with the police, the customs, the health authorities and the pirogue captains, she changed her prize into CFAs. The fake yellow fever certificate did not fool the man in the white coat but he was impressed by the real thing, and the crowded vaccination certificate. It took two hours and an altercation with a man I thought was running off with my passport (it turned out that he was an assistant to the chef du port) to be stamped out of Congo and secure passage upriver: I traded small bribes, cigarettes, submission, conversation, fees and taxes in return for my seat at the back of the motor pirogue. My travelling companions waved from the bank, Bertrand with his arms clasped above his head, as if sending the spirit of victory after me, and the canoe turned into the current. I looked away from everyone, afraid, in a ludicrous way, that I would cry. It was strange and miserable to become companions, to join a group and earn its kindness and feel its protection, and then to snatch

myself away, to go on, always, always on. It was like leaving old friends
for the last time.

A warm heavy mist lifted off the river, trailing soaking fingers
through the trees. The captain was about nineteen but seemed much
older as he read and reread the river, slaloming between up- and
down-currents, tacking up long straits, stitching sandbanks together.
It was a far, far-away place, between worlds, between sense. The
jungle on our port side was Congo but would become Cameroon. Up
ahead of us and to the right what was now Cameroon would become
the Central African Republic. Overhead the sky was thick grey vapour
and around us the world was water, sand and ooze, squeezed between
the roots of bent and fallen trees. An eagle flapped heavily across the
river; I could not make out its species. Ahead the waterway divided as
another river flowed into the Sangha from the west. Down this, at
disproportionate speed, came a blistering white motor boat carrying
eight or so passengers, three of them white men, and all wearing
bright yellow full-body protective suits – not rubberised, like dry-
suits, but light and bulky like chemical warfare gear. The white men
and I stared at each other. One of them nodded, then they were gone.

'Those suits,' I asked the captain, ' – Ebola?'

'No,' he smiled, 'they are loggers.'

'You're sure?'

'Yes,' he said, his smile fading. We nodded through the wash they
had left as if overcoming a bad taste in the mouth. I felt I had not seen
any of my tribe for a year, and then there they were, so insulated from
the river's morning that they made this distant world seem diseased.

CHAPTER 5

The Confines of Cameroon

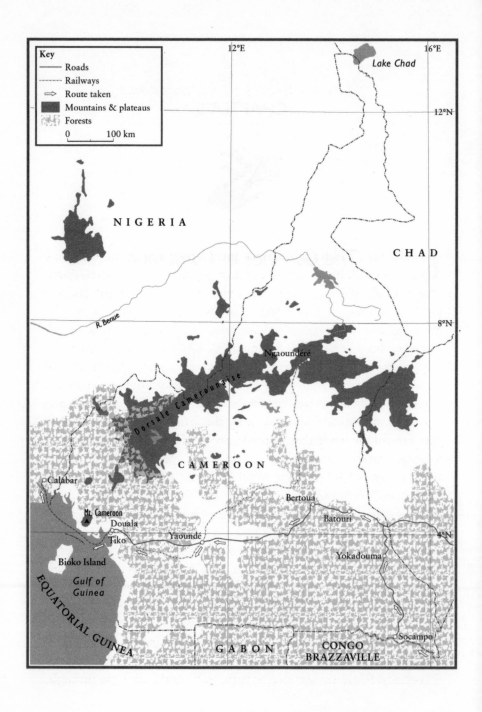

12°E

16°E

Lake Chad

NIGERIA

CHAD

12°N

R. Benue

8°N

Ngaoundéré

Dorsale Camerounaise

CAMEROON

Calabar

Mt. Cameroon

Bertoua

Batouri

4°N

Douala

Tiko

Yaoundé

Yokadouma

Bioko Island

Gulf of
Guinea

EQUATORIAL GUINEA

GABON

CONGO
BRAZZAVILLE

Socampo

The Confines of Cameroon

In Namibia I had feared Zambia, imagining it a poor and desperate place. In Zambia I had feared Congo: the mere word conjured images of slaughter and anarchy, child soldiers and casual death. I knew very little of Cameroon. The police, I believed, would be grasping, and the political situation was unstable: in the last week there had been news of mass demonstrations and riots in Cameroon, sparked by a rise in the price of fuel but swollen by President Paul Biya's suggestion that he might change the constitution to extend his twenty-five years in power. In the last week people had been dying in Yaoundé and Douala, shot down by the army. The guidebook said very little of the world that awaited upriver: 'Cameroon's wild east . . . logging, truckers, prostitutes . . .' was all it made of Yokadouma, a town a day north, through which I would have to pass. Of our destination, Socampo, a frontier village on the river bank, it made no mention. Neither the village nor any road to it appeared on the maps I carried.

I had no food and no knowledge of how the transport networks, if there were such, might operate; I had no contacts or telephone numbers in Cameroon, but I had learned trust. Where there were people there would be food. Where there were roads there would be vehicles. The journey seemed to have a momentum, an onward drive. I was beginning to have faith – in African travel, in the people, above all, but also in myself. As long as I could stay well, and barring any disasters, it felt as though that onrush itself would carry me through.

The first sign of Cameroon was a break in the wet vegetation on the left side, a muddy haul-out and a few pirogues lying in the water like seals. In midstream a large boat with a cabin lay on its side on a mudbank. Up ahead, where the wet sky loured, was a kind of mirage. A truck, yellow as a toy, floated across the dirty silver river on a raft, apparently buoyed on the mist. The pirogue nosed into the muddy slipway and we disembarked. There were three other travellers: a slender man with a blue baseball cap, a young, chubby woman, a hugely broad man with a sports bag and a red baseball cap on backwards. One of the pirogue captains asked me where I was from.

'Portugal?' he exclaimed.

'*Non, Pays de Galles*,' I said again. It seemed a strange conversation to hold, on the slippery interim between the river and Cameroon: though overcast, it was as hot as the brightest Welsh summer day and wet as deepest autumn.

The man looked about him sceptically, as if this was an obvious fabrication.

'*Oui*,' rumbled the big head under the red cap, '*Pays de Galles* – rugby!'

Before we had even reached the customs hut at the top of the slope he and I were talking rapidly, exchanging names of flankers, fly-halves, teams, championships and results.

'I am a rugby player,' he said. 'I have played for Cameroon.'

I never doubted him. Patrice was immensely wide and fit and strong. Aged twenty-nine he had made two appearances for his country, once against Zambia, once against Botswana, playing prop. These are the sturdiest men on the field; in addition to his strength Patrice was also light on his feet, and deft. I knew rugby enough, had watched and played it enough, to imagine he would be formidable.

'I have just been down to Angola,' he said, 'to meet a Portuguese agent. The French wanted me to go to France for a trial but they did not let me.'

'Who didn't?'

'The Cameroon Rugby Union.'

'Why not?'

'I am the wrong tribe,' he said, with a snort.

The customs hut formed one corner of the small muddy rectangle which is Socampo. At the other corners were bars. Nodded heads and pointed fingers urged us up the steps and inside. We took our seats on a bench and the play began.

Behind the desk is *le Gardien de la Paix*, 'the Guardian of the Peace', who looks like a paratrooper, in smock, jump boots, beret and expression of battle-fatigue. Facing him are our party: the Drama Queen, a traveller; Patrice and Blue Hat, also travellers; and me, a tourist. Watching us, seated to the left of the leading man, is the clerk, also in military uniform.

Outside the wooden hut people come and go. Sometimes voices are heard. The midday is sticky hot. Anyone who becomes excited will sweat and reek. An empty chair stands in the middle of the hut, facing the *Gardien*'s desk. It is not clear who is in charge: the clerk, the *Gardien*, or someone else out of sight. Through the open door red mud, green forest, puddles of water and little bar shacks are visible. The *Gardien* gestures me to the chair, picks up my passport and studies it for a long time.

Watching him, my thoughts teem with possibilities. He's never seen a passport before. No, he's never seen a British passport before. (Long pause.) He knows exactly what he is doing. The passport is fine, the visa is fine, I know it's fine, I sweated to get it, it's stamped, dated, paid for, the embassy . . .

The *Gardien* turns a page. Then another.

I decide he is reading the passport, reading it like a Bible. Never did a passport suffer such attentions! Any more pressure on this passport and it will burst into flames! Relax! I tell myself, re-lax . . .

The other travellers shuffle a little. The *Gardien* allows himself a disbelieving shake of his head. He has a broad nose, thick sculpted lips and an expression which now droops with something like amazed disgust. He turns rapidly back to the Congo page, then forward to the Cameroon visa. Behind me I sense the travellers have stilled. At last, the *Gardien* speaks.

'The stamp is out of date. You must pay.'

I jump up, forgetting protocol.

'Look you can see, this should be 2008 but it's 2007.'

'May I see?'

'See – 2007.'

There is an ink stamp, and a little paper stamp, like a star, and – the wrong numeral.

'But I got it in 2008!' I yelp.

'Yes, you got it in 2008, you can see the date here, but you have a 2007 stamp.'

'But I paid for it . . .'

'Yes, you paid for it, you can see the amount here,' the *Gardien* points out, helpfully.

'But . . .'

The *Gardien* speaks, wearily, imperturbably, with the same air of disgust. What is he so disgusted by, I wonder?

'They should have given you a 2008 stamp. To give you a 2007 stamp they should have back-dated the visa so that it ran from 31 December 2007. But' (shrug) 'the stamp is wrong.'

'Fine.'

'There is a charge.'

Boldly, with a sensation that I have suddenly understood something, though I am not sure what, I say 'Fine!' breezily.

The *Gardien* lays my passport aside. I twitch, a gesture which says 'Do you want all my money right now?' and the *Gardien* ignores it and me. He is beckoning Patrice. Patrice rises and hands over his passport. The *Gardien* studies it briefly then picks up the first of three beautiful tools: a rectangular stamp, followed by an oval stamp, then a black biro. The stamps are stabbed into red ink then applied with great delicacy and pushed home hard. There will be no smudge. The biro is flourished. Patrice has not even bothered to sit in the interrogation chair. The passport is returned, he sits and Blue Hat is up next. The same thing happens and he too returns to the bench.

They could both go now, I think, but they wait. We have shared a canoe ride, a short walk and a little gossip, but somehow, already, we are a unit. We are travelling together.

The *Gardien* holds out his hand to the Drama Queen (henceforth 'Queen'). She rises, angling to one side of the desk, and places herself between the *Gardien* and the clerk. She offers the *Gardien* a piece of paper. The *Gardien*'s look of distaste is firmly dug into his lips: now his eyebrows hoop and pucker with incomprehension.

'What's this?'

The Queen mutters something quick and quiet, half in French, half in patois at double speed.

'But it's not you,' the *Gardien* frowns.

'Yes, it isn't because it was taken from me, this is my ID though – I came through here about two weeks ago.'

The *Gardien* looks offended, as if by an evil smell.

'I came through! You stamped me out!' she cries. She is not wasting any time in quiet dispute. She is escalating.

The *Gardien* sighs, shakes his head, closes his eyes and picks up my passport. The Queen withdraws towards the door, muttering.

'It's ten thousand CFA,' he tells me. Not too much. Nobody hisses or clicks behind me – which is a good sign – and anyway there is no choice.

'Fine.'

My passport is painstakingly stamped and returned. It is beautiful. I pay the clerk. The *Gardien* picks up the Queen's piece of paper.

'So this is you?' he says.

'Yes,' she says, surly now.

'But it is not stamped.'

'The one that was stamped . . .' she begins, exasperated.

'Was stolen!' chorus Patrice, Blue Hat and the Queen.

'She had everything . . .' Patrice rumbles.

The *Gardien* raises an eyebrow. There is a slight tightening of lips. The chorus falls silent. 'Stolen?'

The Queen loses her temper again: 'Taken by the fucking police on the other side!'

The *Gardien* studies the pathetic, inadequate, impotent piece of 'ID'.

'But I came through,' the Queen wails. 'You . . .'

'Right!' snaps the *Gardien*, and opens a beautiful ledger like a teacher's marking book, its lines crammed with different coloured inks, the pages crisp and bowed with the weight of biros.

'Which day? What name?'

The Queen circles the desk to stand over him and they both study the ledger.

'Maybe here . . .' she says, her finger yawing vaguely across a page.

'Date?'

'Wednesday, Thursday, ah, the 19th . . . ?'

'Where?'

'Here! That is me.'

'But that is not your name.'

'No, it is.'

'It is not the same name as this ID.'

'Yes because my ID was stolen! I told you – the police took everything. This is my sister's ID.'

'Your sister? Your sister? This is not your ID but that is your name? What the hell is this? A joke?'

The Queen seemed to snap. Her breath came in deep gasps as she rose in a crescendo: 'A joke! A joke to you, not a joke to me! I have been travelling for three weeks and this is what I get every time I meet you, you bastards, I crossed here three weeks ago and you stamped me, you saw me, you put me in your book and then I cross over and they take my money, they take my passport, they take my ID and just to get back to my own home I have to borrow my sister's ID and I had to pay them to get out and now I come back here and you say you can't even find me in your son-of-a-whore book and what can I do? What can I do? What are we supposed to do when all we want to do is live without being robbed and humiliated by the sons-of-whores and thieves who take your money and take your ID and take your passport and write things down for no reason whatsoever and what do they say then – ? Give me your ID and if you haven't got that give me your money! And you have already taken all my money and I have not done ANYTHING WRONG!'

Tears poured down her face. Patrice, Blue Hat and I were stricken by her caterwauling, which swelled to a truly pitiful howl and tailed off in racking sobs. Something like the involuntary wince caused by a baby's screams was carved into all our expressions. Surely, I thought, this must melt his heart. My God, I'll pay whatever it takes to shut her up and get her through. Perhaps we could have a whip-round . . .

The *Gardien* merely looked at her, and at us, as she raved, with an expression of deepening offence. Then he spoke. He did not begin quietly and he rose rapidly to a parade-ground roar. As she had, he played as much to the three of us, sitting on the bench, as to his opponent.

'You have finished? You have quite finished? You are sure? Good, because this is what it is. I have never seen you before. You do not have ID. You tell me a pack of lies about coming through here, but I have nothing and you have nothing which says that is true. You come in here and shout at me and point at a name in the book that could be anybody and then you say we are all thieves and sons-of-whores and you scream about being robbed and just trying to live your life and you have the affront, the nerve, to shout all this at me as if it is my fault? Do you know what I am doing here? Do you know why you are here? Do you know what the law is? The law is my job. That is why I am here, that is why I am asking you for your passport, because it is the law. And do you know what the law says happens now? Let me tell you. The law says there is a prison where they have rooms built for ten men. Do you know how many people are in there? Do you? Thirty-five or forty. Do they care – no, they do not care – they take you and they throw you in, you and ten more like you. And THAT'S THE LAW. So don't you dare come in here and SHOUT AT ME, RIGHT?'

He let us go after that. I do not think he charged her anything. We all withdrew quietly to the bar of the Hotel du Port on the other side of the slipway, where we drank Guinness. I peered into my passport, at the fresh red stamps. Guardian of the Peace Second Class, one said, and there was his beautifully neat, elegant and symmetrical signature across the centre. I cannot forget his expression as he deciphered the tale my passport told. Somewhere, thousands of miles away in London, in the grand white embassy in Kensington that he had never seen, someone was secreting and no doubt selling 2008 visa stamps, which all the records would say had been issued and paid for. No wonder, stuck on a river bank as far from the cosmopolitan world of diplomatic privilege as it was possible to be, he had shaken his head. He looked like a man who was being informed yet again – presented with documentary proof, yet again – of the utter futility of his task. There was something magnificent in his expression of disbelief at the clarity with which he saw the system for what it was: the head-shake of a disgusted and invisible witness, chained to the furthest edge of the absurd scheme of things.

We were 2 litres in before I realised it was 7½ per cent. I read somewhere that Guinness sold more beer in Cameroon than they did in Ireland.

'So how did it go with the Portuguese?'

Patrice's face darkened.

'Oh, he said things. Made promises. Nothing will happen.'

'It was a long way to go.'

'Yes. But I am twenty-nine! For professional rugby I do not have much time.' He cracked his knuckles. Patrice's clothes were very clean, and not cheap. I gathered he was a person of importance in his region, near Bafang, in the north-west of the country: his father was a chief. There was a held-in frustration about him, the bone-deep unhappiness of someone who had been raised to aim high, and blessed with talent, and had had some breaks, and yet – the stars remained beyond his reach. His laughter, his consideration and the generosity and courtesy with which he treated the woman who served us were all the more touching, considering he was in the middle of a 500-hundred mile defeat.

Woozy with booze and rugby chat, we picked our way across the red mud to another bar, where there were two large pans of food. One contained fish cooked in a green stew of something like spinach; the other, cooked in the same stew, held monkey jaw-bones. We plumped for the fish.

The Hotel du Port was a wooden shelter with tables for drinking at, a concrete room with more tables, and behind these, four wooden huts divided into two with a crucifix of corridors between them. We would be taking a truck later that night, Patrice said: we might as well all rent rooms. Blue Hat, a quiet, gentle man called Pascale, agreed. He had travelled this way many times before, he said, buying and selling cocoa in Congo, in a region on the south bank of the Sangha. The truck might not leave for a long time.

The room was a dark wooden crate, infused with damp. There were spiders on the walls, frogs in the corridor and a great many mosquitoes. Out the back, over a wet red mud-bank, there was a washroom and a lavatory, both popular with cockroaches.

Nothing happened in Socampo that afternoon. One or two trucks passed through. Men sat around, ate, drank and wandered off. Patrice forbade me to investigate the bar on the other side of the square. 'Bad', was all he would say.

The grey day faded into a syrupy night. The frogs began a chorus and moths batted themselves against the hurricane lamps. I washed under a bucket, attended by the largest cockroach I had ever seen. The truck would leave at four-thirty in the morning, Patrice said. We met at four-fifteen, under blatting rain, our feet sliding in the mud. The truck would leave at six, we learned. We met at a quarter to six. The truck would leave at half past nine . . .

Socampo woke to another soaking hot morning. Rain in the night had deepened and extended the puddles. Down by the river swallows were hunting; a large group, their swooping and flickering speed made the human world seem almost petrified in comparison, our borders and rituals ludicrous, our travel tortuous. Upriver another yellow truck was pulled across the water on its barge. In front of the customs hut the clerk raised the flag of Cameroon and saluted; the flag hung limply. The *Gardien* was there, hands in pockets, head down, half-smiling as he listened to gossip from an informant. Opposite the Hotel du Port one of the bars was serving coffee and bread.

The truck had a wheel problem. A motorboy changed one of them, bouncing on a long spanner with both feet to break the tension in the nuts. Our bags were flung up to the top of the cargo, Patrice firing them 25 vertical feet with barely a grunt. The lorry was loaded to the sky with crates of empty bottles. We piled into the cab – Pascale, Patrice and I on the broken bench behind the driver; the Drama Queen and an old man with a stick in the front passenger seat.

'How many days to Yaoundé?' I asked again, just to check.

'Three,' Pascale replied.

'We have to be strong, Horace,' Patrice smiled, grimly.

'But we reach Yokadouma tonight?'

'Yes, if we are lucky.'

'I do this every month,' Pascale said. The truck's gears crashed and we began to climb another incline. The road towered and dipped, twisted and turned; the forest here rose and fell on waves of hills. As we ran down slopes the bottles behind us leapt and bounced, swayed and concertinaed with a spectacular noise that threatened to turn into avalanches of breaking glass. At the bottoms of valleys we steamed forward as fast as possible, making a run at the next slope which we would top at walking pace. Sometimes we ran out of power: then the driver gunned the engine, whipped through the gears, sweated and cursed: at such times his effort and dexterity and will were all that rowed us on.

The bench at the back of the cab was falling apart: iron struts stabbed up through torn foam rubber. We balanced between them as best we could, as tense as jockeys.

'Every month?'

'Yes – twice!'

'You must be crazy!'

'Ah, but I have a wife and child.'

We stopped in villages to take on sacks, passengers and more charred pig: all this went up on top of the crates. The passengers were mostly boys and young men riding from village to village.

'How is the cocoa business?'

'Good. And there is gold.'

'Gold?'

'Yes, I can show you – give me the map . . . see – here, there is gold here.'

'Right – do many people know about it?'

'Yes. Some. More will come.'

'Give me some of that water there,' Patrice boomed.

We came to a truck on its side. It was one of the older machines, nothing more than a cab and a long drive-shaft stretching to the back wheels. Above the drive-shaft were the tree trunks. The driver had taken a nasty right-hand bend on a down slope too quickly. Now he and his motorboy were sitting in the shade under the trunks, and grinning. We offered them a lift but they declined, seemingly relieved to be off the road.

Our truck was an old Mercedes, blue, battered and tireless but very slow. The vast majority of the traffic was composed of the yellow logging trucks which rumbled through the forest in an irregular stream. We pulled over for them: they must have been twice as heavy as us, but they went faster. They were made by Renault, either brand-new or in excellent condition, and each of them carried three or four or sometimes only two massive bodies of felled trees. The trunks were extraordinarily straight and a beautiful dark red-brown. In the medieval world of the forest, where the villages were little collections of huts, animals, people and cooking fires, the yellow trucks were like spacecraft: the unbelievably long arm of a rich and distant planet, reaching into this misty world of laterite red and dark glowing green, efficiently plucking its treasure.

As in the Republic of Congo and the Central African Republic, a dictator has been a huge benefit to the European powers doing business with Cameroon. French, Belgian and Italian logging companies dominate the plunder of the forest: France is the largest importer of tropical hardwoods in the European Union, with a 28 per cent grip on a trade estimated at €500 million a year and almost certainly worth much more. The logging companies receive direct and indirect support from the French government and meet no opposition in the wilds of Cameroon: once they have secured a concession, the forest is at their mercy. The loggers are particularly fond of salvage permits, which were originally designed to allow for the sale of trees which stood in the way of road construction. Officially outlawed for a decade, the permits are still widely used.

Roads which were no more than trenches of mud arced off to either side of ours at regular intervals, many strewn with the trunks of small trees, apparently considered worthless. These tracks went nowhere and served no purpose, except that down them the bodies of great and valuable trees had been dragged. One of the leading French tropical timber companies, Rougier, admitted to procuring over 25,000 cubic metres of hardwood from this eastern region in 2004, five years after salvage permits of the kind it operated were made illegal. The logging companies might be forgiven for regarding legality as merely another surmountable obstacle: one, Vicwood-Thanry, a French-operated, Hong Kong-owned conglomerate, happily pays fines of half a million dollars a year: a kind of formalised bribe to international law.

The fruits of the plundered forest, one of the last primary rainforests on earth, may be seen sawed, stacked and stamped on the docks of most major European ports. They can be bought in DIY chains like Monsieur Bricolage, and admired in refitted kitchens all over Europe, Japan and China.

'Women!' Patrice rumbled, with surprise. We were coming into Lokomo, something between a village and a town composed of a long line of huts and shacks, a sawmill and a roadblock. I strained to see over the driver's shoulder. Between Patrice's biceps and the wall of the cab, barely a foot wide, my perch allowed for little movement. There was singing and a great deal of whistling. The sun poured down on a huge crowd, all in the same brown kikoys, speckled with yellow, red and white, and all female.

'International Women's Day,' supplied the Drama Queen. She was a resourceful traveller; I had barely noticed buying her lunch (fish and manioc). The truck stopped and we climbed down, sighing as the blood flowed into our limbs, and stumbling slightly as the carnival engulfed us. Shouting, singing, cheering, whistling, dancing and, above all, drinking, the town was entirely in the hands of the women. They arrived in trucks, shouting and waving, and departed in trucks, still shouting and now waving bottles. They stormed the bars, eddied around the trucks, and clapped and stamped and swayed.

The truck needed another wheel-change. The driver went to eat, the motorboy fetched the spanner and jack, Patrice and I were nominated to source the beer.

'Which bar?' I pondered.

'That one?' Patrice seemed equally tentative.

'You ready for this?'

'*Oui!*'

'OK, don't be scared, I'll be your bodyguard.'

From then on Patrice delighted in introducing me to everyone we met as his '*garde du corps*'.

We laughed and puffed and squeaked in protest as we were pawed and wrestled, seized and danced with, pointed at and laughed into the bar, then out of it. With Pascale we formed a small tortoise around the motorboy who gulped his beer, wide-eyed, and rushed the wheel-change. The driver gobbled his meal and we took to the truck like spooked dogs, rattling out of town to a roaring chorus of leering women's shouts.

I climbed up onto the crates, unable to stand the crushing and stabbing of the bench any longer. The motorboy and his friend were amused: would I ride with them on the roof of the cab? I dared not, and made myself as comfortable as possible in the very middle of the top layer where our bags were stacked. Notwithstanding the flexing of the crates and the unsettling way the bottles below jumped up into my buttocks at every significant bump, it was an improvement, rocking along at a third of the height of the forest, staring into the kaleido-scopes of leaves, blooms and branches. In villages I was spotted and given cheers and waves: it seemed extraordinary that in the hundreds of kilometres we travelled, that day and the next, I never saw another white person, and not one of hundreds of locals showed me anything other than friendship. The logging, the chaotic but persistent destruction of their environment, was being perpetrated by European companies to feed the appetites of Europeans like me, but there was a simple visual cut-out between the planners and beneficiaries, and the victims. Apart from that glimpse of three white heads on the speed-boat there was no sign that the outside world had anything at all to do with the pillage.

The sun sank and the stars came out. We began a long descent which entirely collapsed what little orientation I had. It seemed to go on for hour after hour: how high had we been, that we could still be descending? I watched the stars through the high trees as planes of the world seemed to tip beneath us. Sometimes it seemed the truck must surely overbalance, heels over head like one of those slinky springs that can be made to tumble downstairs. The motorboy crawled over to join me for cigarettes. We accumulated more and more passengers, and more sacks, as we neared Yokadouma. They did not seem to pay anything; it was as though the beer truck was part goods-train, part public transport. Sometimes a group of boys sang as we went, a lovely sound which disintegrated into laughter when a particular lurch made us all grab our handholds tight.

We arrive in Yokadouma not far short of midnight on 8 March. I only know the date because it was International Women's Day. Dates would have been entirely irrelevant had I not had to hit certain windows according to my visas – Algeria's in particular. We descend cautiously from the truck, watched by a quizzical woman cooking over a charcoal fire, and find our feet on the broken pavement.

The Drama Queen says she has an arrangement and vanishes on the back of a motorbike. The roads, which are wide and sloped, darken swiftly as we move away from patchy lights. Everything is made of mud: the road, the pavements, the walls of houses, the floors of the hotels we found. The first place is too awful, even for Pascale, Patrice and I in our numbness. We settle for the second, a sticky, clammy roach motel. The walls weep and things scuttle away from the light. We find beer in a smashed bar – everything is broken, spilled and upturned, as if it has just hosted a monumental fight – and then go to look for food, cursing the motorbikes which slice out of the darkness rattling like gunfire, climb pavements and rip up to the front of bars. There are only men and excited boys: perhaps all the women have gone to Lokomo for International Women's Day. There is an angry, worked-up atmosphere; drunks lurch out of the darkness and shout things I do not understand. Yokadouma seems to be a function of the logging business: illegitimate, unloved and unbalanced. A little way back up the road, still lethal with truck-traffic, exhaust floating spectrally in the red glow of rear lights, we find men eating and drinking at a single crammed table. Patrice orders for us.

'*Donnez-moi des oeufffs là!*' he commands, like a small Etna, throwing out an indicative finger on the end of one huge, imperative hand. We drink milky sweet coffee and eat golden eggs, chewy bread, oil, salt and processed French cheese. Tearing into the bread I feel a click in my mouth and a large section of one of my canines breaks off and falls into my mouth, leaving a raw socket. Oh hell, I think, spare me a Cameroonian dentist. But there is no real pain, so I spit the tooth into the gutter and carry on with the meal. Later we drag ourselves back to the hotel and a damp oblivion. The bus is at first light tomorrow.

'*Ça va*, Horace?' Patrice mutters, as we lug ourselves over the mud to the bus station in a sullen dawn.

'*Bien sûr! Quelle bonne matinée!*'

We laugh and slap hands. Today will be our hardest. We will be crammed in rows of six into a little blue Renault bus the shape of a small loaf of bread. It will have no springs. When the sun comes out it will become very hot. The road will mostly be ridges, teeth-clashing, head-banging ridges that go on and on and on. I will get a window again and pay for it with a constant pressure from the body beside me, driving my ribs into the side of the bus: I will collect a fat bar code of bruises down my right side. We will travel out of the forest, into woodlands. We will be stopped at a dozen roadblocks. We will have to bribe one, and the entire bus-load of us will be involved in an argument at another on behalf of a young man whose identity papers will be questioned. I will feed a stream of rehydration sachets to a small girl, travelling with her father, who has a sinister grey diarrhoea. Three children will travel with us in a day-long odyssey, impossibly harsh by European standards, at which none will complain even once. I will see two groups of swallows, heading into a north-westerly breeze.

In mid-afternoon we arrive in Batouri, where the driver and our most important passenger, a fat army officer whose presence has melted the roadblocks, burst into roars of laughter at the sight of my face, coated in the orange-red dust of the road. I try to shift the red coating with a backyard bucket, suspiciously observed by chickens, while Patrice and Pascale order fish – a dorade, served with mayonnaise and a small cloud of envious flies. The plates were washed in a bowl of brown water, the cook's fingers were thick with fat and stained with ash and the flies were so persistent that they barely left the fish for an instant when you flapped a hand at them. We eat with our fingers, with relish. I have no fear for my stomach: I will eat anything put before me now, regardless of the conditions of its preparation. All trace of my fastidiousness has gone. Partly because I am so thoroughly inoculated, partly because nothing can be as bad as Ebola, and partly because what little food we come across is delicious. The dorade is cooked to perfection.

In the evening we change bus in Bertoua. The composition of the travellers shifts: many are Muslims, all are men. We set off at higher speed into the gathering dark. First the road runs through forest, then at last the shuddering ceases and we hit tarmac. We stop once more, beside a long bright strip of shops and bars, where gas lights, bright hand-painted signs and activity of people eating, drinking and milling around take on an hallucinatory quality: we shuffle around yawning until the driver herds us back into the Toyota where we reassemble into a drowsy heap, heads on each other's shoulders, feet under the armpits of men in front and half wake, half doze as a large moon rises, the lights disappear, reappear, flash and thicken, and then roads, and houses, and at last, at midnight, Yaoundé gathers around us. It is huge and shapeless, with many more buildings than lights. It seems to have no centre, no focus – dark streets, shacks, hills covered in lightless buildings. I am suddenly grateful I am travelling with friends. It would be no fun to arrive here alone, exhausted, late at night.

Pascale is going on to Douala, a journey which will take the rest of the night. His endurance is spectacular; we have all found a place in our heads where we can take refuge, leaving our bodies swaying and bumping to the irregular rhythm of the road, but for Pascale to do this journey twice a month must take another level of steel. We swap email addresses and embrace.

'Don't take the train to the north,' he says, and Patrice agrees vehemently.

'Why not?'

'*Petits voleurs!*' they chorus together: Little thieves.

Patrice and I take a taxi. Tonight is on me, I insist: Patrice knows a hotel. It is grotty but cheap – tomorrow we will find a better one, he says.

'Goodnight then,' I bid him.

'You are going to bed?'

'Yes! Aren't you?'

'*Non!* I am going out. Come and drink!'

'Oh God. I don't know . . . Let's clean up. Call for me when you're ready.'

The bathroom has the first mirror I have seen for days. My face and hair are dark red and my features and body have changed shape. My face used to be rounded but is now triangular. The cheek bones are prominent, the sockets of my eyes have lost their softness. There is a stillness in my gaze, either of exhaustion or of ease. I am as unworried, these days, as I have ever been. My body is pared down. Over my flanks, hips and stomach there is no spare flesh at all. I am over a stone lighter than usual, I reckon. The travelling seems to have altered the way my body works. I have had none of the stomach or bowel problems I believed were guaranteed for a European on the road in Africa: on the contrary, what food we have eaten seems to be absorbed directly into my system, leaving almost no waste. I can go without food for a day and not really think about it. I can fall asleep instantly and wake to an unfamiliar wakefulness immediately. I can sit on a bus, my eyes open, with the rest of my body almost asleep, in neutral, almost oblivious to discomfort, armoured against the passage of uncomfortable time with a blank, bovine patience. Europeans have no idea of the routine physical discomfort of travel as people travel here. You never have a seat to yourself. The heat is omnipresent and unrelieved. And everything always breaks down.

Even my walk has changed: now I am very light-footed, immediately aware of any imbalance of weight in my rucksack. By adjusting it, and my stride, I can move the pressure around my feet: no more blisters.

'Sebago . . .' Patrice had said, admiringly, looking at my feet.

'What?'

'Sebago!' he said again, pointing at my shoes. Other people had said the same thing; only now do I understand. There are little ribbons on the sides of my shoes, showing the make.

'That's good?'

'*Oui!*' Patrice nodded, emphatically – 'The best!'

There is no reason why it should seem so strange that here, of all places, people should be label-conscious and brand-savvy, but it does.

I take a shower, which runs red, then a cold bath, which is also red. I photograph it and my face with my mobile phone.

'Let's go!'

Patrice looks very smart. We go out into hot, dark rain. There is very little traffic moving in Yaoundé.

'Where are we going?'

'Club.'

'What club?'

'You will see . . .'

We sample two. The first has an empty restaurant downstairs; upstairs it has torn-up plush, bright lights and extremely loud music. Behind the bar the server is asleep: Patrice wakes her gently. Onstage seven women are dancing. They turn their backs to the audience, bend over to reveal buttocks under short skirts and begin to shake. They shake and shake, rotating, humping, jerking and shaking again, all their faces set in concentration as they study their own reflections in a mirror at the back of the stage.

The second club has very low ceilings and red lights. The customers are a mixture of women and men, many in couples. The entertainment is stripping, only the stripping is done out of sight: a tall, naked woman emerges and makes her way through the throng, stopping in front of couples to gyrate and thrust. The couples giggle wildly. She approaches us. I am so tired that this is all becoming a hypnogogic dream.

'Please,' I say, weakly, as she begins to jerk her groin towards my face, advancing with tigerish steps.

'Please, madame,' I try again, 'I am so tired, really, I am just here for a drink . . . please, I implore you, excuse me!'

I look at Patrice. His face is set in a monumentally stern expression, as if we are about to take on an implacable foe in battle. If I had the energy I would laugh wildly and run out of the club but my body has given up. The naked woman turns on Patrice who sets himself, elbows on knees, fists together and stares fixedly ahead. She fills his vision with her midriff. The music crescendos and Patrice does not blink. It is single combat now; Patrice's granite gaze versus the woman's most intimate part, which she thrusts towards his nose.

Patrice does not quail. The music climaxes and stops. There are cheers and the whole club applauds – the woman, primarily, but also, one senses, Patrice. I have seen men stare down a few things, but never anything like that.

Yaoundé is built on hills. On their crests and ridges are the skeletons of buildings like the ruins of Rome. But these are not the relics of antiquity; they are the shells of a future that never came to pass. Concrete frames, with doorways and holes for windows, their construction was halted when the money ran out. Now they stand, unsafe, rain-weakened, waiting for someone who will not come to pull them down and start again. Around their feet are the shacks and bungalows of the city. We switch hotel to a relatively homely place overlooking the railway tracks. All day and all night there is activity down there as trains shunt, load logs and clank sombrely away down the line to Douala, the port and the world across the sea. It seems to be the only thing that works and never stops working in Yaoundé, this mechanism for transferring tree trunks from trucks to flat cars.

Through television, the newspapers and gossip the political situation becomes clearer. I stand with groups of silent men, when the evening editions of the papers come out, staring at the headlines on news-stands. The demonstrations lasted for three days. In riots which began as a transport strike over the fuel prices maybe forty people died, maybe more. The authorities used tear gas, water cannon and live ammunition. Now many of those arrested are coming to trial and being rapidly sentenced to prison. Newspaper editors have been threatened and radio stations taken off the air. As a last word on all this, the government has just announced the budget for the next twelve months: the police and the army have been given significant pay-rises.

'That is all it takes,' says a barman. 'The army are the best-paid in the country. They will shoot us to protect their salaries. Nothing is going to change.'

Patrice comes into my room and sits down. He is awkward, tense with a kind of frustration I have noticed in him, as if his huge muscles crave something solid to test themselves against. He takes off his cap and spins it unhappily in his hands.

'Horace, will you help me?'

'Of course. What do you want?'

'You know I am a rugby player, and I am good.'

'Yes.'

'But they would not let me go to France, to play.'

'No.'

'So will you invite me to Wales?'

'Of course! But why do you want to come?'

'Because if I can get a coaching certificate from a European Rugby

Union I will be able to work here. Even if I cannot play, I can coach. If I could visit you I could get a certificate and come back . . .'

'I got one of those at school. It was easy . . . I don't see why we shouldn't get you one. But how will you pay for your flight?'

'I have a friend . . .'

Mr Kenneth is much older than Patrice, small and bald with a direct gaze. He is a businessman based in Yaoundé for whom Patrice had done a little work – mostly lifting and packing, as far as I can tell. I am prepared to be suspicious of him, as he is of me. We speak English, Mr Kenneth being an Anglophone, and interrogate each other, politely, as Patrice watches us, head turning from one to the other as he tries to follow our conversation.

'But why do you want to pay for Patrice?' I ask.

Mr Kenneth nods rapidly as he makes each point.

'Because I have seen him play rugby and I know that he is good, because there is nothing for him here, because I do not want to see him waste his life. If we could give him a chance . . .'

I was certain of him very quickly but we talked for half an hour and then I summed up.

'I cannot promise anything but I will do everything I can to help Patrice come to Britain. I have no house of my own but when he arrives in London he can stay with my father and me for a night or two and then we will go to Wales where my mother lives. She has a farm and spare rooms. I am sure she will be very happy to put us up for a couple of weeks, and Patrice could certainly help us with the farm. In Wales we will try to get Patrice a coaching certificate. I have friends who are into rugby there: if we can, we will get Patrice a trial, or at least introduce him to a Welsh club. I cannot promise he will get a coaching certificate, but we will try. At least he will see something of the world, and I will repay him for his great kindness to me.'

Mr Kenneth nods gravely.

'I will pay for Patrice's flight,' he says, 'and I will give him some money for Britain.'

We look at Patrice like two fathers who have hit upon a plan for the advancement of a child. There is something about his bulk, his directness and his troubled-son's air of having been helped and of needing help that inspires one's urge to help. It is as if he was not made for the world Mr Kenneth and I know, of complications, of probable failures, of connections, strings, snakes and ladders, but for the parallel, brutally sanitised world of the rugby field. How amazing it would be to welcome Patrice to Britain, to take him to a rugby game, to get him a certificate – and what a fairy tale it would be to see him play, to 'put his body on the line' as they say in the game, for the pleasure of spectators; how magical it would be to give him a shot at the kind of triumph his great hero had achieved: the Cameroon-born flanker of the French national team, Serge Betsen. We shake hands and embrace. Patrice goes to his room and comes back:

'Horace, I have something for you,' he says, and presses something into my hand.

'I am also a sculptor,' he says, shyly. 'This is how I make some money.'

It is a small bronze bull with buffalo horns. It is very heavy and very beautiful. It stands on three legs, the fourth being slightly raised, as if pawing the ground. The power of the bull and its barely arrested forward charge are eloquently present in the metal. Greatly moved I hug Patrice again, then we go out to see about the visa: how hard can it be?

'My God, he's only a bloody Taff!' booms the white-haired man. Three nights later I am having a gin and tonic in the hotel bar: the white-haired man has been gossiping with one of the waiters. His French is very fluent and the waiters all seem to adore him. One of them has just asked me where I am from.

'How do you do?'

We introduce ourselves. His name is Barry and he is British.

'I am in the oil business,' he confesses, not embarrassed, but confessional nonetheless. 'I have been to every single country on this continent.'

'How does Cameroon compare?'

'Bloody terrible! Terrible! Haven't you seen it?'

'Well, a little.'

'Do you know – the people here have coins? Fucking coins! Not even notes. Coins. Do you know what that means?'

'No.'

'It means they've got nothing. It means that when they die there is a ditch, a bloody ditch, which they throw your body into.'

'What can they do?'

'They should have a revolution, that's what. I've been telling them that for years. I've even told the government! But nothing changes.'

'What can we do?'

'Oh, you try to help. You know, I've been helping someone for years. She wanted to sell SIM cards. You've seen all the kids selling SIM cards? Well, I tried to help her do that. But then someone in her family got ill and she gave them all the money. Then she wanted to sell something else, and so I helped her set up again. It didn't work, all the money vanished, you know, bills, family. The government gives them absolutely fuck-all. Even schools aren't free; as for health-care, forget it. If you get ill, that's it, you're probably dead, and they throw you in the ditch.'

To apply for a visa for Britain from Cameroon you first have to log into a website. Despite Cameroon being a Francophone country, this website, and the visa form, are only available in English. Guessing that the best way of getting the process right is to speak to a human being, we try the British Consulate. The lady at the gate turns us firmly away: 'You must apply through the website,' she insists.

We download copies of the form at an internet café and take them back to the hotel. Patrice needs a letter of invitation, proof of income, an itinerary and something to prove that his accommodation in London actually exists: I email my father and arrange for a copy of one of his utility bills to be couriered to Mr Kenneth. We go over the form several times. Though a trial with a rugby club would be a clear infringement, as the holders of tourist visas are not allowed to seek employment, I can see nothing to suggest that gaining a coaching certificate would be against the law. There is a box on the form in which anyone who has helped with the form must identify themselves, which I do.

In the evenings we pick our way along broken pavements to small cafés, where we are served minuscule portions of thin, delicious chicken with manioc or rice. Patrice devours them in seconds. My stomach has shrunk: one small meal in the evening, with a breakfast of eggs the next morning is all it seems to require.

On the fourth day – visa application being a slow business – Patrice receives a telephone call. I watch his face as he takes it. He looks stricken.

'My wife is ill,' he says, 'malaria. I must go home.'

'Horace . . .' he begins. A black face blushes as readily as a white one. I press a mixture of euros and dollars into his hand and we dash to the offices of Western Union: Isabel will get anti-malarial drugs this evening. We make our arrangements on the run to the bus station. When my father's utility bill turns up Patrice will be ready to submit his application. He will let me know by email when he has done so and I will track its progress by calling the consulate – they are likely to want to speak to me, we believe.

We hug. I lift Patrice off the ground.

'*Force, Patrice!*' I cry.

'*Force, Horace!*' he returns.

We shake hands.

'See you in London,' we say, and he goes.

The Nigerian Embassy in Yaoundé is much quieter and more efficient than the one in London. Two visits, a long wait in a Chinese restaurant across the road, gawping at huge bats which circle and flap in broad daylight like flocks of gulls, a third visit and a short sleep in reception yield a visa. I have given up on taxis now, being short of money and depressed by the regularity with which they break down, or sit in dammed streams of smoking traffic. I walk everywhere. When I am nervous I carry my notebook and pen: a sword and shield, I tell myself, as I did in Lusaka, then in Brazzaville. No threat appears.

There are two routes to choose from: north-east to Maroua in the marshes south of Lake Chad, west to Maiduguri, then Kano in Nigeria and north again to Zinder in Niger. I study this and dismiss it. There has been fighting in Chad, with French forces backing the government against a rebel insurgency. The swallow route, based on the last time I saw a great many of them, should lie to the west, nearer the coast. That is the way ornithologists say they come south – across the Gulf of Guinea. I will take the train to Douala on the coast, then look for a boat across the armpit of the Gulf, up into Cross River State, in Nigeria. From there I will go due north, across Nigeria and into Niger, striking west again to meet the Niger River at Niamey. I will go on to Agades, way up in the Sahara in northern Niger, not far from the Algerian border, where I will use a number that Christine in Brazzaville gave me for a guide named Akly. There are rumours of Touareg rebellion on that border, but if anyone can get me through to Tamanrasset in Algeria, she said, he could.

The Douala train is drawn by an old diesel locomotive which looks like a steel egg-carton stuck on top of a shoe-box. The carriages have fewer doors than doorways – in the last, where I sit, the back is open, showing the tracks tailing gently away. We roll out through slums, into hilly wooded country, over river beds, through a morning which heats and turns smoky yellow. At the first stop, all around the train, suddenly, whipping in swirling dives above the roof of the station, there they are: a flight of swallows.

I rush to the open door to stare at them. There are twenty or thirty; they come and go too quickly for me to count them. Their moult looks complete: their new blue backs shine in the hazy sunlight like hardened silk. It seems extraordinary and somehow unbelievably simple: they have kept to the course I plotted from the Mambili; they have overflown the whole of the rainforest, crossed Gabon, and now here they are, feeding ferociously. I see more flocks in Makak and Eseka: I am on to them again, I have found them. The day becomes hotter and hotter and the carriage more crowded, as much with hawkers who travel from stop to stop as with travellers bound for Douala. The train seems to lose speed rather than gain it but I am perfectly happy. This is the most luxurious, gentle travel, and I am keeping pace with the birds. What I am seeing, they see. Where they have been, I have been. Where I am going, I dare hope, they are going; I feel a drowsy oneness with them and fend off sleep by speculating on the effort it costs them to cover ground at the speed I am, slumped in the seat. Four wing-beats a second, 240 a minute, 14,400 an hour: suppose they fly for eight hours a day, at a conservative guess – over 115,000 wing-beats. It seems ludicrous, and perhaps it is. Along with swifts and martins, swallows are the smallest birds capable of gliding flight, which assists in hunting, and even more so with migration. Without the ability to glide, their task would be akin to riding a bicycle over 6,000 miles while prohibited to free-wheel. Swallows have a much higher power to weight ratio than most birds thanks to their wing-loading: at just over 14 grams per 100 square centimetres this is much lower than that of most birds. The bird's shape, the swell of the breast tapering to nothing at the very end of the tail streamers, is similar to a racing yacht.

For a long time the elongated tail in male swallows was cited by scientists as a classic case of a male trait exaggerated by female choice, but recent studies have suggested the aerodynamic effect of the streamers' ability to twist when the tail is spread allows a long-tailed male to avoid stalling in a high angle of attack, and also generates lift which aids manoeuvrability in slower, turning flight. Intriguingly, studies show that at their summering grounds in the north, where mating and breeding take place, male tails are on average 10 milli- metres longer than the optimum length for flight – and it is well known that females select long-tailed males. The swallow's pattern of moulting in the south and regrowing their streamers as they travel north thus forms a rather wonderful trade-off between the demands of migration and the demands of mating. When they most need their extraordinary powers of flight, the tail is at its best length to provide them. When they most need a seductively long tail, it is ready.

The outskirts of Douala are rough and ridden with smoke. The heat pounds the train, radiating out of a sky the colour of new corrugated iron. The train moans slowly across points.

At the station I make phone calls and establish that there is a boat to Calabar in Nigeria, which leaves tonight.

'Where does it go from?'

'Tiko. You must buy a ticket before five o'clock,' says the woman on the end of the line.

'How do I get to Tiko?'

'There are buses, but – you do not have much time.'

It is gone three o'clock already. I agree a fare with a taxi driver and we set out.

'No trouble,' he says. 'It does not take too long.'

We circle the hem of Douala, lurching through heavy traffic. It is much busier and much hotter than Yaoundé.

'Oh yes,' says the driver, 'this is the economic capital; here everyone comes for money.'

There are mills and the sweet smells of sawn wood. There are lagoons in which there are no birds, there are glimpses of ships and the roads are jammed with people, buses, cars, scooters and lorries. We drive along the edge of one market as crammed and chaotic as anything I saw in Brazzaville: only the pot-holes are not quite as fierce. We stop briefly to get a permit allowing the taxi to leave town, and stop again to have the permit checked. Then we are out onto a motorway. On either side, dense groves of thin grey trees all lean seawards and a familiar smell fills the car, the thin, tightly clean smell of a bicycle shop: we are driving through a huge rubber plantation.

The taxi breaks down, steam boiling under the bonnet. The driver fills up a water bottle from the plantation's irrigation system, which is intricate and hugely extensive: someone has a lot of money invested in these trees. It is something of a shock to be among them, to smell

them; this is the odour of the stuff so many millions died for, that
made so many thousands rich. And yet this is not the same plant that
made King Leopold over a billon dollars, that baubled-up Brussels
and seeded so many villains' villas on the French Riviera: that was the
rubber vine, which grew wild in Congo. This is the rubber tree,
imported from South America and first widely propagated in this
region after 1908, when Leopold was finally persuaded to hand over
his fiefdom to the Belgian government. Here, in what was French
Equatorial Africa, the rubber tree served French companies well,
thanks to a Leopoldine system of hostage-taking, forced labour and
brutal punishment. In 1911 competition from plantations in South
America and Malaysia (largely under the control of Britain) caused a
crash in rubber prices. Companies operating in French Equatorial
Africa lost profits, so cut their expenditure. Transport and collection
networks reaching into the interior were abandoned: it was
plantations on the coast, like this one, which continued to operate.

I could not prove it but there was no reason why the men crossing
heavily, heads down, from one tract of trees to the next, and the trees
themselves, should not be descendants of the people and the plants
which were here a century ago. Only now, thanks to the hissing
irrigation system, fewer men are required per hectare – each of which,
it seemed reasonable to assume, supports even more trees.

It is night in Tiko on the coast of Cameroon and hot under a smoky,
black-brown sky. The moon is a sardonic yellow grin – 4° north of the
equator it fills from the bottom, like a wineglass. The waiting room is
a concrete shed, thick with heat and the smell of bodies. A fan makes
no impression but the noise from the television reaches all of us. I am
beginning to hate that television: tuned to a Cameroonian channel, it
is showing one of the interminable soaps in which people are always
weeping by hospital beds, finding each other dead on floors, over-
acting like mad and agonising about who will tell the matriarch the
bad news this time. I have watched a lot of television, in various hotel
rooms, and nothing has struck me as distinctive except a soap I saw in

South Africa, set in an office and apparently focused on the lives of a bunch of gay Afrikaners in striped shirts.

I wake from a half-doze on the hard bench to shouts: 'Let's go! Let's go!' Everyone leaps up in a flurry, grabbing their bags and children. We are herded out of the back of the shed into a yard where a huge lorry is waiting. It has four long benches, with room for twenty people on each, and is constructed like a giant crate with rectangles cut into the side. Light comes from a bulb swinging dimly above our heads. We clamber in with maximum panic, which does not augur well for our voyage. When we are all seated the truck fails to start – and fails again. Everyone sweats and tuts. I have picked up the habit of teeth-sucking to express disapproval. When eighty people do it at once the lorry suckles and bubbles with the sound of hissed saliva, as though we are a load of consumptive crickets. Teeth-sucking is wonderfully expressive. When stopped by a policeman, as we so often were in Cameroon, the whole bus sucked – ffsscch! (meaning 'Bloody cops'). When a tyre burst, as every tyre seemed to, sooner or later, we sucked: wwssccchh! ('Here we go again'). When the bus did not turn up, or the truck did not leave – mmwwch ('Well, there you go'). When a ticket seemed expensive, when an hotel was full, when a room was grotty, mwvcchh . . . ('We'll just have to live with it, but we're unimpressed'). It came so readily to me now, and was so viscerally, accurately expressive of feeling, I worried about doing it inappropriately: My wife is ill – mmmwwchh . . . ('Well, that's bad but what can you do?'). The words it saves are countless, and impossible for the hearer to misinterpret.

After half an hour of sweating in the semi-dark the engine begins to turn over, unconvincingly. A man addresses me.

'Hey, White Man, are you a mechanic? Can I see your qualifications?'

'No, Black Man, I am not. Who are you – the police? Can I see your ID?'

The woman opposite laughs at this, throwing her head back and jabbing her sack further into my groin. The truck starts. We bounce through Tiko, my stomach complaining about supper, which was rice with a lump of gristle and a short tubular section of rubbery artery. We pass a checkpoint and enter the port area, a dark swamp about 2 miles wide. The engine gives up after half a mile. There is more sucking of teeth, more sweating and more banging from the driver as he punishes the machine for its failure. The mosquitoes feast on us. Everyone piles out, then scrambles back in as the engine fires. We bump through the swamp to lights, shiny black water, two fishing boats, and ours. It is not a big boat, about twice as long and twice as high as the fishing boats, and it is entirely covered with people. It is Dutch-made and rusty. In the bowels is a canteen where a crowd is piling into spicy chicken on skewers, and beer. Lifejackets are stacked around the walls and all the seats are full. God help anyone down here if anything goes wrong. In the queue for food I meet Mark, a tall Cameroonian about my age. He is extremely quick and funny: we buy beers and withdraw to the upper deck to watch the rest of the loading.

Mark speaks like this: 'Heeeey, maaan! Have you ever had African pussy, maaaaan?'

'Well . . .'

'Where do you live, maaan?'

'In London, some of the time –'

'I live in Seoul in Korea. I teach English to Korean kids – you should see them! They're good!'

'Oh yeah?'

'And I own a hairdressing salon. I make about 3,000 US dollars a month, maaan – do you want to see a picture of my wife?'

'Sure.'

'Look, here's a picture of my car. God, I love my car, maaan – look, there it is, and my wife, in the snow. It's such a great car – it's so cheap to run. You know what I've been shipping CD players and things to Cameroon and because they actually work I'm making real money, maaan! There's another picture of my car . . .'

'Very nice.'

'It's so easy to make money out there – and I love the attitude – not like Cameroon!'

'I guess things are harder here.'

'Huh! No good blaming the government, maaan! That's all people do here, sit around and complain! They're lazy! They don't have to do anything! The food grows on trees! They want to make the country better they gotta get up, stop complaining, and make it better!'

'But it's a dictatorship and the army will shoot anyone who tries to change it, won't they?'

'True. But, well, I'm gonna make money in Korea then come back and help my country, because I love my country, maaan!'

For complicated reasons which I struggle to follow, Mark is going to Nigeria to renew a visa, or fulfil the conditions of a visa which will allow him to continue to commute between Korea and Cameroon.

The boat is being loaded with sacks of a leaf which looks like cocoa, and bags of an unidentifiable vegetable. These are rolled off trucks and bundled into the stern. The loaders are working quickly but not fast enough for the boat crew: an order is given and the front ropes are cast off. Every flat surface of the deck and superstructure is covered in men and women. Many are eating now; the journey is out of their hands for a while, and they are relaxing. There is shouting at the back, two more sacks are tumbled onto the deck, and the stern is untied. We move off, into dark channels. The helmsman switches on a powerful searchlight which probes ahead of us, roaming across dark masses of mangrove. It is eleven at night and still the sweat runs off us. The boat twists and turns, following the arm of the searchlight through the mangroves. Now large speakers on the rear passenger deck begin pumping out an Africanised hip-hop beat: it feels more as though we are cruising around the block trying to turn girls' heads than setting sail for the Bight of Biafra and the Gulf of Guinea. Then a preacher starts up, bellowing over the music, haranguing the travellers who are packed together on benches under an awning – how is it, he wants to know, that God's miracle has been proved for all to see? He has two accomplices. After a good harangue they bless us and we all join in with the Amens. The preacher prays fervently for our safety on our travels and we all join in with the Amens, even more fervently.

'I went on a boat from Korea to China and it was nothing like this!' Mark says. He looks rather shocked.

Now a breeze picks up; a warm fluting of air around our sticky faces. We turn into the open sea. The boat puts on a gentle roll, which becomes a less gentle corkscrew. Looking through the window of the bridge Mark and I are relieved to see there is a radar, which is working. Suddenly everyone feels tired. Mark goes off to find some deck to sleep on; I lie down where I am, on the wing of the bridge, on the seaward side. Lightning flashes behind the clouds to landward. We are passing between Bioko island, part of Equatorial Guinea, and Mount Cameroon, but both are invisible. There are no lights, the moon has vanished and the stars are blotted out by thick cloud.

CHAPTER 6

Nigeria: Gulf of Oil, Coast of Slaves

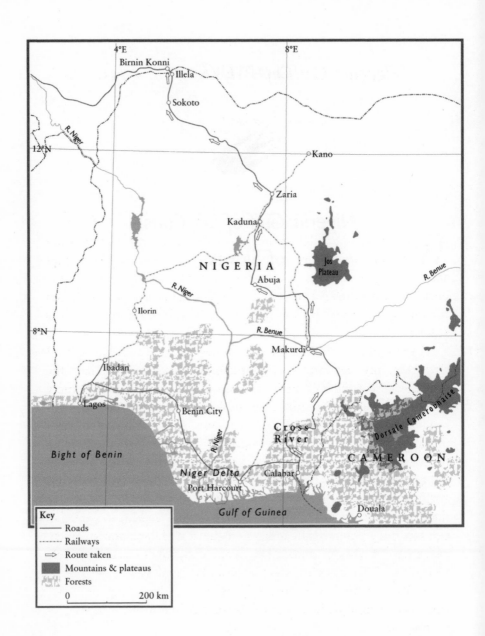

Nigeria: Gulf of Oil, Coast of Slaves

We wake to rain falling quite hard with a hissing noise. Bodies are stirring around me, cursing quietly as those of us without shelter struggle and slither, crab-like, towards sheltered parts of the deck which are already over-full. I fall asleep in a sort of crouch. When the rain stops I move back to my former spot. It is wet but the air is still hot.

At three in the morning the rain comes again, heavier this time. I struggle to my feet, drunk with fatigue. The land is invisible in the dark but to seaward half the horizon is on fire. Oil flares, rigs, supply ships and tankers form a motorway of light. In the depths of the night it is business as usual in the Gulf of Guinea, one of the hottest oil spots on earth. I stand there, swaying, staring at the illuminated sea, trying to distinguish between different ships and rigs, breaking the strings of light and fire into separate vessels, too tired to care about the rain and wishing I had some emergency whisky with which to fend it off. The door to the bridge opens and a figure beckons me in. It seems an extraordinary kindness: I can barely believe it. Thank you so much, *merci, merci*, I whisper, and curl up, making myself as small as possible in a corner. I black out and sleep deep and dreamless.

Daylight comes with lots of ankles and legs, engine noise, conversation and the luxury of the hard, dry floor of the bridge.

'*Bonjour!*'

Heads turn, faces look down from above: '*Bonjour!*' they grin. I stand, feeling excellent, as though all my muscles have been wrung out.

'There you are!' says Mark. We return to our spot, yawning. He found a good place to sleep too, he says. A man comes around with milky coffee and a sweet pastry, which, strangely, has some sort of fish paste in the middle.

We are cruising up a wide channel of mangroves and palms. It looks like an ancient, eternal Africa of picture books, of *Just So Stories* and legends. The palms with their heads nodding slightly to one side and the mangroves, clutching the sea between their roots, seem to promise fever, snakes and slavers. Here and there, through breaks in the green walls, are villages made of reeds, half-floating. In the channel canoes are working: they are heavy wooden things, flat like big punts; some are being paddled, some sailed and one or two have motors. There are three men in each, throwing out and hauling in nets. The channel is thick with flotsam: lumps of wood and clumps of mangrove bob out to sea as large terns dive into the water, fishing, and supply ships steam past us, heading out for the rigs in the gulf.

'What is the absolute proof of God's mercy?'

As it gets hotter the preacher starts up again. His assistants come round handing out leaflets, then the Bible, instructing us to read a bit and then pass it on. Around half past ten the channel turns right into a large river, and there is Calabar.

There are houses along a ridge, warehouses and sheds by the water, naval vessels and wrecked boats half-sunk, and other wrecked boats so battered and rusty that they really ought to be sunk but which are in fact being loaded, and a wharf crammed with a crowd of hundreds of people. Over it all, against a hot silver sky, a huge green and white flag of Nigeria, the biggest flag I have seen, flaps slightly.

The door of the captain's cabin opens and the captain emerges. It is the first we have seen of him; we must have been in the hands of the mate. The captain looks fat and self-important in gold-rimmed shades. As children burst into tears and loud women fuss over them the captain takes over and noses us towards our berth. We edge between two rusty freighters, and all seems well until the captain forgets about our stern and it swings viciously. There are cries of alarm on the other boat, men run, and ropes stretched between one of the freighters and the shore first take then fail to hold the tremendous pressure of our weight. They drag the freighter towards us and there is a mighty crash.

Disembarking means charging down a narrow plank off the nose of the boat into a heaving, howling, shouting, waving, surging throng of people. I am the only white, as at least a hundred people have noticed and a good dozen point out. Dozens more want to shake hands, say hello and find out where you have come from. I keep smiling and keep bullocking in. I am rapidly adopted by a young man who leads me in a kind of race around the immigration people, who stamp my passport, and the money-changers, who have stacks of naira which they manipulate at dizzying speed (a crowd of nodding faces assure you the exchange rate is fair), to the medical people.

'You have a yellow fever certificate?'

'Yes, it's in the bottom of my bag.'

'Your bag?'

'Yes, here.'

'OK, don't worry about it.'

The next thing I know I am on the back of a motorbike and ripping up the hill, out of the port and into the razzle-dazzle of Calabar's permanent rush hour. In Nigeria, I surmise, they do not hang about.

'Who are you – Simon Peter or one of the other disciples?' demands Josephine, the receptionist at the hotel. She offers to show me around town when her shift ends in the afternoon, on the condition that I shave thoroughly and change. We set off in her little red Volkswagen.

'My father gave me this car as a graduation present,' she giggles. 'I am still not very good.'

'You are just fine,' I gasp, soothingly, as we career around a roundabout. We go first to a salad bar, which is deserted but for a waiter watching a soap. Someone weeps by a hospital bed as Josephine picks over a huge salad of tinned vegetables.

'I live with my brothers, but I am so fed up with them, they do not do any cooking! So I get up, cook their lunch, go to work, go home and cook their supper. My new boyfriend has a good job at a bank. Perhaps I will marry him and leave my brothers – ha ha!'

Josephine says her job is a bore. She has graduated in Tourism and Hotel Management and now wants to move to a city, but as yet she does not have the money. It is a shock, after Congo, Cameroon and Zambia, where jobs like hers are in the hands of older people, mostly men, to find a young woman evincing exactly the same kind of dissatisfaction you would expect to find in a European graduate in a similar role: it is as though in crossing from Cameroon an older century, almost another continent, has slipped away. Josephine wears a crucifix and although she does not go to church she does believe in God, and in the tenets of the Church.

'Do you have a wife?'

'No.'

'Are you homosexual?'

'No. Why are you whispering?'

'Because here if you say you are homosexual they will stone you!'

'Don't be ridiculous! Why would they do a thing like that?'

'Because it's a sin!'

'Of course it's not! Stoning people definitely is, though.'

'Yes it is – homosexuality is against God.'

'No it isn't, if you believe God makes us who we are then you have to believe He makes some of us gay.'

'But it's a sin!'

'No it isn't. I know vicars who do God's work who are gay.'

She almost falls off her chair.

'Do you want to see the museum? I have not seen it myself yet.'

'Yes please.'

We swerve and stall our way to the river, a little way downstream from the port. There is a deserted visitors' centre, a line of empty cafés and a museum. Bad music plays from speakers attached to lamp-posts and vultures circle the car park. It is nearly closing time but the duty guide says there is just time for a tour. We take the fastest tour of

Calabar's museum of slavery that there has ever been. We hustle along a dark spiral corridor, pausing in front of exhibits which light up automatically. They are composed of animated life-size figures in tableau, with recorded voice-overs.

A woman begs for mercy as slavers seize her from her village. Men moan in agony as they are shipped across the sea. Americans laugh heartlessly and bid for slaves at an auction. A slave howls as plantation-owners beat him.

Astonishingly, the museum's narrative makes the Americans the ultimate baddies of slavery. The British, who thanks to the heroic Royal Navy put a stop to it all, come out, lauded with a chorus of 'Amazing Grace', symbolised by portraits of Wilberforce and John Henry Adams, as the best of eggs. Whoever had set out the museum took great pains over context. Slavery, it is pointed out, existed all over Africa for centuries before the Europeans became involved. Africans enslaved Africans, Arabs enslaved Africans, then Europeans came and everything culminated in the plantations of Virginia and the Caribbean, where villainous masters made fortunes out of misery.

But the central point of Calabar's history is that between 1720 and 1830 over a million people were here loaded into little boats, crammed head to foot in decks less than 12 inches high, and taken out to sea where the deep-draught slave ships were waiting. Between 1700 and 1830 Calabar was effectively twinned with Bristol in the most horrifying manner imaginable – Bristol-based merchants held a near-monopoly on the trade through Calabar. During the period of British involvement on this coast, 'the Slave Coast', otherwise known, in a linguistic twist far beyond irony as 'the White Man's Grave', over two million people were transported from West Africa to the New World. A third of a million died en route.

With 'Amazing Grace' echoing in our ears Josephine and I sat at a row of empty tables, ordered fizzy drinks and stared at the grey-brown water and the green swamps of the river's edge.

'This is where they took the ancestors away,' Josephine said, quietly.

Small bands of swallows came up the river bank, feeding easily as they went. I felt sick and hangdog, with a kind of guilty squirming in my stomach. The river banks, the mangroves, the still palms and the birds were all here, exactly like this, when my people were enslaving Josephine's. Guilt, particularly the guilt of the white man in Africa, is a useless, much-mocked emotion: what a bleeding heart, to sit at a café table and lament what cannot be undone. Furthermore, Josephine's actual ancestors had not been slaves, or she would not be here in Nigeria, and mine were not slavers. So what I felt, I realise now, and I can still recall it, as physical a sensation as a shiver, was a confusion of guilt with shame.

'How is God's mercy manifest? What is the absolute proof of God's power?' the preacher demands.

We are the last Abuja-bound travellers to leave Calabar the next morning, packed into a Toyota minibus which is promisingly well maintained and functional, but we are not going anywhere until the preacher has finished with us.

We listen dutifully as he explains that our very existence, all existence, indeed, is proof of God's mercy, and that our continued survival is absolute proof of His power. We pray that we will have His protection on our travels this morning, this afternoon, tomorrow and ever more. We pray, we say Amen, we are blessed, we say Amen again, we pray again, until everyone has been through a restive and uncomfortable period, and emerged into a numbed – or enlightened – acquiescence, whereupon there are final Amens, the preacher withdraws and the driver takes his seat.

It is a little after nine in the morning. The traffic in Calabar is heavy, as usual, but at least it keeps our pace down. Emerging onto the main road to Aba the peril becomes apparent. Nigeria has excellent roads, the best I have seen since Namibia: the problem is this allows drivers to proceed at the maximum speeds their vehicles can summon – and they do. Our driver is a fat, impassive man with a look of frozen discontent, who drives us like the clappers. After the first five miles I consciously force my stomach to relax, my fingers to loosen and my chest to breathe normally, and commit my soul to God. At least the end will be quick, judging by the blackened wrecks at the sides of the road.

Cross River State, through which we are passing, seems to be in the middle of a low-level guerrilla war. Occasionally bites of this conflict appear in the western media, particularly incidents involving kidnapped or shot-up western oil workers, which therefore affect the oil market. The morning newspapers are full of all the rest of the battle: police machine-gunned, rebels shot, bombs thrown, houses burned, people displaced, soldiers ambushed, guerrillas killed, arrested, tried, imprisoned. The three most common sights on the road this morning are soldiers, gas stations and churches. It is a close thing, but I estimate that the churches dominate.

The sky is a hot kind of dusty yellow. Fumes belch from the lorries and I marvel at the churches as we hurtle between them like a sort of jet-propelled unholy spirit. The Chapel of Glory, the City of Life, SuperChristian Church International, Last Days Christian Fellowship, the Unified Church of God, Jesus Family Church, the Established Church of Christ . . . Many are no more than a small square building; some seem to be merely a roadside sign.

We stop to fill up with gas. I have been sitting in the second row of seats: directly in front of me is the poor young man who has had the best view of our multiple near-death experiences. He is shaky on his feet.

'What do you do?'

'I am a model,' he says. 'I am going to Abuja for a shoot.'

'How's it going?'

'My career is going well,' he says. 'Do you have a cigarette?'

The problem is finding somewhere to light them. The road is lined with gas stations in both directions. The air is heady with the smell of petrol.

Nigeria is the United States of Africa, I think, amazed by the volume of traffic on the roads, by the swiftness of transactions, by a pace and a beat which seem everywhere stronger, louder and more pressing. The clientele in the café where we stop for lunch are unlike any similar gathering of travellers in the places I have passed. The men wear smart shirts and good shoes; the women are still in all the colours of Africa but now these are cut and stitched into the styles of Europe. Labels and gold, fake and not, line the long table. Meals are picked at fastidiously; since South Africa I cannot remember seeing food being thrown away.

The day changes colour as we go north. We cross the Benue River, a dark orange tributary of the Niger rolling sluggishly through low mud-banks under smoky copper clouds. The land feels almost unearthly under the sky's strange light. God and Gasoline, I keep thinking: the two most powerful, most dangerous genies in the modern world are here both so present, so ubiquitous, it feels as though a single spark could ignite the very air.

The afternoon newspapers are gripping thrillers. Here are terrible stories of women being circumcised by their in-laws, men being shot for their car keys, former ministers making off with hundreds of millions of naira and freely confessing it, state governors attacking each other and their predecessors, youth groups turning on politicians, rumours about sects within the military planning coups, committees at war with assemblies, and public-spirited reassurances that lovers who have gum disease should not worry about kissing their beloveds because gum disease is not transferable. The newspapers eat up hundreds of high-speed miles as we hurtle towards Abuja, an invented capital city.

Nigeria was a constellation of kingdoms, peoples and religions. In the north was Bornu, on the western side of Lake Chad, and the trading cities of the Hausa. Bornu was Muslim by the eleventh century, as were the Hausa traders by the fourteenth. The Fulani people adopted a stricter Islam than was practised by the Hausa, and prosecuted it through war. By 1809 they ruled what is now northern Nigeria from the desert town of Sokoto. West of the Niger were the lands of the Yoruba, and the states of Ife and Oyo. In the forest south of Oyo was Benin, whose leaders were still using human sacrifice in the nineteenth century.

With the arrival of the British, first as enthusiastic slavers then as equally enthusiastic anti-slavers (evangelising palm oil instead), came first trading stations, then consuls, then conquerors. Lagos was taken in 1851 and annexed to Britain in 1861. By 1900, having partly burned down Benin City in a reprisal raid, Britain ruled everything from the Niger delta in the south to Bornu and Sokoto on the edge of the northern desert – and the constellation became, in the eyes of Europe at least, one vast territory.

Britain's administration of these regions between 1900 and independence in 1960 develops as a tortuous series of reclassifications, with administrators struggling to impose unified rule on diverse peoples and kingdoms. A federal structure is created which facilitates 'Indirect Rule', in which powerful chiefs rule small fiefdoms under British supervision. In 1914, two blocks of fiefdoms, hitherto known as the Protectorate of Northern Nigeria and the Niger Coast Protectorate are merged into a single entity: the Colony and Protectorate of Nigeria, a simplification which causes more complications. The northern, eastern and western regions each have their own Houses of Assembly by 1951; by 1954 there have been three new constitutions in eight years; by 1959 all three regions have achieved internal self-government and Lagos is a federal territory, like Washington DC, home to a federal government dominated by northerners.

Independence in 1960 creates a nation of 115 million people, divided into 200 tribes, speaking over 500 languages, living in a federation in which the north is predominantly Muslim, the south Christian and the whole riven by internal conflicts. A rebellion in 1966 by the Ibo people of the east leads to Ibos living in the north being massacred. The following year the East declares itself the independent Republic of Biafra. The Biafran War lasts three years and ends with a famine and Biafra's surrender. Since then, dictatorships, coups, swindles, flawed elections, riots and inter-ethnic strife have failed to prevent the country being declared the happiest in the world by the *New Scientist* magazine, which claimed to have surveyed the happiness of sixty-five nations and found that Nigeria came top.

Perhaps the most concrete consequence of the country's short, vexed and often bloody post-colonial history is Abuja, a city created in the early 1970s as an alternative to Lagos, symbolically situated in the middle of the country, and, perhaps also symbolically, designed by a consortium of three American firms. It was built in the 1980s and became the national capital in 1991. We rocketed into it, as the late afternoon began to turn to evening, at top speed. The edges of the road eddied and swirled with dense crowds of people. Huge shanty towns have grown up outside Abuja in recent years: roadside cooking, crabby buses and jostling crowds made a narrower and narrower passage for the increasing volume of traffic.

There did not seem to be anything to Abuja – away from the packed suburbs the centre was a network of huge, wide, near-empty roads. At the bus station I shook hands with my model friend and jumped into a taxi.

'Please take me to the biggest hotel you can think of, where they are most likely to have satellite television,' I begged the driver. It was Saturday 15 March and Wales, having beaten England, Scotland, Ireland and Italy, was about to play France in Cardiff for the Grand Slam.

I missed the game but was included in the celebrations by text message. Thanks also to text messages, I made contact with an old friend whom I had not seen since school. Fifteen years ago she was a quiet and thoughtful half-Nigerian, half-German student of International Relations. I did not know what she was doing with her life now, but she seemed to have become rather grand: 'I will send you my driver,' her message ran. 'He will be in a Jeep, registration . . .'

The jeep and the driver arrive and whisk me through the dark and deserted boulevards of the city. We pause, turn off the highway and pass through a high steel gate, which closes behind us.

'What is this place?'

'This is the French Compound,' the driver replies.

Low bungalows, a school, a swimming pool and a clubhouse are threaded through with trees and lawns. There are soft lights illuminating pathways; from the pool area music plays.

The driver knocks on the kitchen door of one of the bungalows, and there she is. It is like being seized by a whirlwind.

'Look at you! Not an ounce of fat!' she cries, as we hug. She was always strong, a dancer, and she is even stronger now: a European girl has become an African woman.

'Have you eaten? Would you like a drink? As you can see I tend to eat and work at the same time . . .' The television is tuned to CNN and faces her place on the sofa, in front of the laptop. 'I will switch it off now,' she says. 'Come on!'

At the clubhouse she is the only non-white. People are surprised and delighted to see her.

'I never come out here,' she says, sidelong, 'except to swim. Do you want to swim? I really need to . . .'

The moon is high behind gaseous clouds and it is still hot. Before we swim I manage to sit her down for a few moments. She speaks clearly and very quickly, her dark eyes locked on mine, her hands telling half the story.

'OK so what the Growing Business Foundation does is micro-credit, partly, and loans for small businesses, and we push for improved policy frameworks, because half the problem is policy. At the moment I have this idea about providing computers for community centres for kids after school, and I am trying to help nomads, because you know in the Sahel they are just . . .'

The Sahel is the semi-desert region where the Sahara shades into a more habitable area, where vegetation begins to overcome the sands. It is one of the most vulnerable environments on earth to rainfall, or lack of rainfall. Traditionally only nomadic farmers could make a living there: the ground could not sustain settled agriculture. With increases in population, the fragile ecology of the Sahel has come under fatal pressure: over-grazing of scarce grasslands and tree-felling for firewood, together with more years of failed or insufficient rain, are effectively expanding the desert southwards.

We swim. She hurtles up and down like an athlete, then we are out and back in her house, eating.

'I fly business class now,' she says. 'And I move people. If I am sitting next to someone who can help, then great. If not, I say excuse me, would you mind, and I swap them for someone else.'

'What are you doing tomorrow?'

'I am flying down to the delta to talk to the Ogoni, then back to Lagos for a meeting.'

She does not say it so simply, but it appears that her foundation raises money from banks, oil companies and big business and puts it into people's hands at the lowest level. She talks about doing this sort of business in Nigeria: about seeing the price of assistance before it is offered with its price disguised. She talks about watching powerful men torn between aspiration and temptation. My head swims at the merest intimation of the complexities of tribal, political, historical, corporate and social networks which she continually processes, cross-references and engages. She talks about the disillusionment of the disenfranchised and the poor, about the enthusiastic corruption of the rich, and about a new employee of hers who has just been to a village and returned, lit up with inspiration at the possibility of being able to contribute to its people.

'It's a strange place to live,' she says, looking around the compound the next morning, as if she has not really looked at it before. 'But my girls like it when they come home from school, they have friends, and they can roam around . . .'

She hugs me again and her driver takes me away. I think of her, working, with the laptop under one hand and the plate of food under another, the lights on, the television on, the telephones on, doing three things at once and in the back of her mind always counting, counting the days until her girls come back from England for the school holidays. She fills every week, every night and every weekend with work. The Nigerian media have already identified her as an inspirational figure: perhaps some day the world will know her name.

Abuja has two significant seasons, but this is the third, between the wet and the dry: this is the season of the harmattan, the wind from the north-east, which fills the air with a yellow haze of sand.

Nothing much moves; one or two kestrels, and straggled flights of swallows – the first time I have seen them in a city. They are not pausing, but going north. I am hurried too: according to my Algerian visa I should be entering that country in four days, and I still have to cross Niger.

Today I will cross the fault-line that runs across Nigeria, dividing the Muslim north from the Christian south. Beyond that line Niger, Algeria, Mali, Mauritania and Morocco are all Muslim. Below it, behind me, is Christian and Animist Africa. A frontier in many ways far more mighty and significant than all the borders I have crossed will pass, unmarked and invisible. God is about to change His name, the demands he places on His followers and the list of His prohibitions. As symbols, the swallows are about to change too. Associated with the resurrection in Christian Europe (because they arrive at Easter), with household gods in pre-Christian Rome and with good fortune in ancient Greece, in the Islamic world they are also seen as a holy bird. The Koran tells of Allah's sending swallows to defend the faithful, when an army of Abyssinian Christians is besieging Mecca:

'And he sent against them swallows in flocks; claystones did he hurl down at them.'

Some writers have speculated that the idea of the swallows hurling stones may have arisen from a mistranslation: rather than 'stone' the original sense may have been 'smallpox scab', implying that the swallows spread disease among the Christian army. Although there is no biological record of swallows carrying disease to people, in our own time there have been numerous scares about bird-borne diseases. West Nile Virus and Avian Flu both summon images of birds dropping from the sky, dead, bringing plagues against which our frontiers, our lines of control and all the demarcations with which we divide the world, are powerless.

White Peugeot taxis are the way to go north to Sokoto. We are cramped again, but not nearly as badly as I have been. We are driven too fast again, but yesterday has raised my terror threshold. The road sweeps out of Abuja and the scenery rapidly changes: first the greens become duller and more tired, then the petrol tankers are replaced on the roads by lorries carrying tottering stacks of firewood, all heading north. As the Sahel becomes denuded and the Sahara spreads south this most ancient form of fuel is trucked up to the pasturalists of the north.

The ground is yellower and hotter and the towns become whiter. Now instead of churches we pass mosques. Changing taxis at Kaduna, where riots over a Miss World competition killed 150 in 2002, a woman begs for food or money – the first time I have been asked for either since Congo. We travel all day. Strangely, though the sun sinks in the afternoon the temperature does not abate. The truck-stops and taxi parks are chaotic, crammed and deafening, but away from the towns the country seems to stretch out and empty. We come to Sokoto in the dark. The hotel is hot and stuffy; in the room next door a large party of Muslims ignore the Prophet's injunctions against alcohol and play loud warbling music. I have a hotel routine, now. Check for hot water. Seal all cockroaches in the bathroom. Strip bed and check for insects. Run through all the television channels, right to the end of the sixty-ninth screen of fuzz. Watch one cycle of TV5 Monde, the French rolling news channel. Write diary, if not too tired (in which case do it over breakfast), otherwise watch a film. Lying on a bed on the Nigeria–Niger border I watch Mark Wahlberg in the true story of an out-of-work, divorced man who becomes the star of an American Football team. The film makes great play of the hardship of living in a depressed, post-industrial American city, but it fails to make it unattractive: people stick together; men play football; there is a pretty barmaid interested in Mark Wahlberg, who gets a job in the same bar. Philadelphia looks fantastically advanced compared to Sokoto or even Abuja. The camera's polarising filter makes colours richer than they are; the casting director has chosen an attractive supporting cast. The hardship of western life is made to seem seductive, like the first part of a fairy tale destined to end happily. The films which come closest to showing our lives as they really are are never worldwide block-busters. It is as though we cannot help but flaunt our fortune in the face of the rest of the world, or, alternatively, that we cannot quite bring ourselves to be honest about it.

The next morning another crammed truck-stop yielded a comically cool driver and a battered Nissan.

'Where are you going?'

'Niger.'

'Right let's go!'

'But I want to share with other travellers.'

'Ha! No one else is going to Niger – what's there, after all?'

'Oh OK then . . .'

A dusty landscape turned into a sandy one, with thorn scrub, acacia trees, and camels. If there was still a distinction between the Sahel and the Sahara I could no longer see it.

We stopped a little way short of the frontier at a roadblock. The driver performed a U-turn.

'Goodbye,' he said.

'Any ideas?'

'Look for a boy with a motorbike.'

The boy appeared. We rode for a while, then turned off the road at the border, a collection of huts scattered like litter around some thorn trees. My motorcyclist whirred me from one to another. Neither he nor any of the other locals crossing the frontier was asked for anything by the border guards, which gave the disorienting and peculiar impression that a frontier invented by Europeans was still manned and maintained primarily for their use. A young man who said he was a member of Niger's security police beckoned me into a hut where the fan had died and the table was not safe to put weight on. We were about the same age, equally bemused, and addressed ourselves to the task of completing various documents as though it was a joint test, which we passed, to mutual satisfaction, in record time.

The pale grey heat of the morning deepened and the sky became a hot, naked blue. The roads of the little town of Birni n'Konni were soft sand and the atmosphere was entirely changed and yet familiar, suddenly, as if the breath-held tension of Nigeria was released, in crossing the border, into an endless sigh. People moved slowly,

keeping to the shade, drank coffee and spoke French again. I sat at the
perfumed feet of a money-changer who explained that all the Central
African CFA francs I had brought with me from Cameroon were
useless: here in Niger they use the West African CFA, and no one
would change one to the other. The bus station was a quiet sandy
compound, a shaded waiting area and one or two other passengers,
including a young jewellery trader who showed me silver from the
desert and talked about rebellion.

'The Touareg are at war,' he said.

The boys selling fizzy drinks and boiled eggs waited in the shade,
emerging reluctantly into the flaying sun when someone approached
their stall. The bus appeared, on time, and we boarded. Most of the
passengers huddled in their seats, holding ragged curtains across the
windows against the light. My neighbour was a Touareg, an elderly
man whose dark brown face was beautifully framed by his blue robes
and turban. We compensated for having no common language with
elaborate politeness. We never took a sip of anything without offering
it to the other; he stood aside for me with great show; I thanked him
with bows and smiles.

Beyond the window the desert was made of light overlapping in
unfinished rectangles: silver-yellow sand ran to a sky which leached
into a sun too hot, too bright, too vast to look at. Shadows were small,
dark living things; everything else was still. The thorn trees, the
scattered dark stones and the emptiness around them were motionless. Road-workers and swaddled men hanging on to passing trucks
waved, moved and made noise, but under the rule of the heat and the
intense light the midday seemed indifferent to life, if not hostile.

We stopped every couple of hours in little villages of colour and
shadow, where clay walls made dwellings and yards conjoin,
simplicity multiplying into intricacy: a hut was linked to another via a
compound wall which in turn was connected by a passage to a
mosque. It was as if the villages were seed-beds for growing and
propagating shade. As we pulled in, people rushed forward: women
half-sold, half-begged; boys waved bottles; men had furls of mutton
skin and fat, frying on hot plates, and long knives with which they

pared pieces off. Many of our passengers disappeared, at a sundown stop, to pray.

At one stop we all disembarked to eat and relieve ourselves. The lavatories were a row of low huts at the far end of the compound where the coach had halted. Those in need hurried towards them, then, as they approached, veered sideways away from the doors. The stench was stomach-turning. I followed the stream of passengers around to the back of the building, where a narrow strip of ground between the shed and the fence was strewn with faeces, with people squatting among huge lizards which seemed to be feeding off the waste. It was the most revolting latrine I had ever seen. I'll hold it for as long as it takes, I resolved.

As the sun sank and the sky trembled with carmine and purple streaks, the driver turned on his radio and a high, racing song came from the speakers: a keening sound, fiercer than a lament and more urgent, driven by drums. My neighbour saw me lift my head to it and he smiled and said the only words of his I understood: 'Rai!' he said. 'Algerie . . .'

Rai is a form of protest music from Algeria, as culturally powerful and pervasive as hip-hop is in the West. The drums race, the melody is fast and breathless and vocals are a kind of bitter, longing wail. I could not make out many of the words of this song except the chorus line: '*C'est payant, Monsieur, c'est payant . . .*'

'It costs, sir, it costs . . .'

While hip-hop has spawned a global industry in which enormous profits are made from ever-higher production values, rai is composed of the caustic, sometimes scornful, anthems of an entire generation of young Algerian men who grew up squeezed between violence, corruption and poverty. Rai does not sell perfumes, jeans or trainers.

As the desert darkened, the sound seemed to swell. It was like a secular mirror of the summons of the muezzin, but it was neither a call to prayer, nor to arms or celebration; it was more like a sardonic lament. While hip-hop's central motif is machismo, the aggression of young men, their totemic evocation of guns, gangs and girls and their rage against the ghetto, rai sounds like the product of a sadder, wiser culture. It is not the sound of an oppressed section of society, but of entire societies, entire countries, ghettoised; it evokes archipelagos of the dispossessed.

CHAPTER 7

Niger: A Quiet Little War

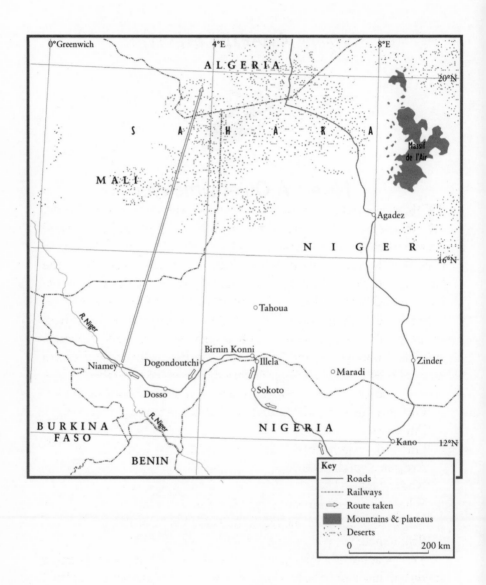

Niger: A Quiet Little War

The canoe leaks steadily: we pass around the cotton and the knife, taking turns at tearing off strands from the grey-white lump and packing them into the cracked wood with the blade. While one of us does this the other two paddle. Niamey soon fades back from the banks and the town becomes the country: low sandy shores crumble into the ruddy river. In some places the edges of the water are alive with people washing, swimming, splashing or fishing; between them are peaceful stretches peopled with kingfishers, geese, ducks, herons and ibis. Sundown on the Niger is a busy time. Our captain knows many of those we pass: fishermen coming downstream on their canoes wave or nod gravely. Our captain is a person of some importance.

'Giri-giri,' he says, with a wink, as if this explains it.

'I'm sorry?'

'Giri-giri – witchcraft!'

'Are you a witch doctor?'

'Not any more. I was.'

'What happened? I mean, what changed?'

'Ah, well, people became frightened, so I had to stop.'

'Frightened?'

'Yes! Imagine, if a man comes to attack you and he comes running at you and just before he hits you he raises his stick, his club, and Paff! The stick in his hand turns into powder! That happened to me, and people began to say I was giri-giri, so I stopped.'

'Do you miss it?'

'No. And my wife is happy that I stopped – ha ha!'

'Is there much giri-giri in Niger?'

'Oh yes, a great deal. Your swallows, for example.'

'What about them?'

'You can use swallows for witchcraft.'

'How?'

'Well first you must catch a swallow. That is very hard to do. But you must catch one and kill it – with a sling, or with gum on the reeds where they roost. But it is very hard.'

'Right.'

'You get a pot and heat some water on the fire and you add some special herbs.'

'OK . . .'

'Then you add some more herbs and heat the water for a long time, stirring and stirring . . .'

'Yes . . . ?'

'And then you put in the swallow, not the lungs, but the heart and everything else, and a tiny bit of oil, and heat it for a very long time, stirring and stirring . . .'

'Uh-huh . . .'

'Until it makes a kind of paste. Then you eat it and you will be protected from car accidents.'

'Car accidents?'

'Yes. And plane accidents.'

'Plane accidents!'

'Yes! Imagine you are in a plane and something crashes – bang! The plane goes down. But not you. When the plane crashes you will not be there.'

'Where will I be?'

'You will be standing not far away, on the ground. Zap! Like that! If you are in anything, a train, a truck, a plane, a car and it crashes – bang! Zap! You will not be in the crash, but you will be near by, if you have eaten the swallow paste.'

'Wow!'

'Yes. Protected. But only for five years.'

'Five years? Then what?'

'Then you must be very careful.'

After a while we turned downstream again. One or two swallows had passed over in the early morning, and now one or two more came up-river, but they were few compared to what I had hoped for. Staring at the map I had imagined the Niger would form a great swallow highway, running from the Gulf of Guinea north into the heart of the Sahara. Perhaps they did not need the river: the birds do use watercourses – I had seen them flying up the Rhône in great numbers one spring – but in good weather, with insects abundant, it seemed likely that they were either fanning out, feeding as they went, or had already taken to higher altitudes, where they might find tail winds to help them across the Sahara. I scanned the skies in vain. My timing was still good – I had not fallen behind the main body of the migration, as far I could calculate it – but my positioning was out. A great many swallows must already have been engaged in the desert crossing.

According to the maps the single greatest obstacle to a European-breeding swallow's migration is the Sahara, but the statistics contradict this impression. Storms, collisions with traffic, predators and most of all sudden and severe changes in weather patterns are the principal killers of swallows: the greatest desert on earth, it seems, presents a surmountable challenge.

Before they fly south in autumn the birds build up fat reserves to fuel the crossing of the Mediterranean and the desert: those flying down the Italian peninsula, which make the longest sea crossing, add between 30 and 40 per cent to their lean body weight; those flying down through Spain add slightly less. The heaviest birds will set out weighing around 24 grams, around 4½ grams of which will be fat. It has been calculated that those 4½ grams are sufficient to fuel up to 1,600 kilometres of flight – allowing the biggest birds, incredibly, to cross the Sahara and the Mediterranean in one long-distance flight, without stopping.

The birds also put on weight before the migration up from the south, adding the same 2 to 4 grams of fat. To cross the sands they must fly non-stop for fourteen to sixteen hours a day: there are few recorded sightings of them in the desert itself. The cost of this effort is clear in a study of swallows arriving in southern Europe in 2002: on average they weighed 13.4 grams and were carrying only ½ gram of fat. This would still have afforded them 200 kilometres of flight, and beyond that, like other migrant birds, swallows have an emergency reserve: when fat reserves are close to expended they are able to break down proteins from the breast muscles and gut.

According to another study, the keys to survival for large numbers of north-bound birds lie in the weather, vegetation and insect populations of North Africa, especially Algeria. While the sea and sand crossings come early in the south-bound migration, when the birds are carrying their maximum fat reserves, on the north-bound leg these two obstacles come at the other end of the journey – especially for the furthest-flying, most northerly breeding birds. Only after several thousand miles of Africa do they reach the Sahara. Conditions in the rich coastal crescent beyond the sands are therefore crucial: here many will feed and build up strength for the push into the various weathers of the European spring.

My first sight of Niamey, that night, coming in on the bus, was of a low-built, dimly lit town, which even after eleven at night still seemed to pulse with heat. On the first morning I rose at dawn and scanned the skies for swallows. I crossed the John F. Kennedy Bridge, one of Niamey's few landmarks, where camels were as much part of the rush hour as lorries, and peered through my binoculars at downstream islands of rice paddies, gum trees and reeds. Very few swallows came over. In the heat of the day I began to research the next stage of the pursuit. Everything was an effort; the heat drained strength first, then will, then concern. Walking 50 yards out of the shade seemed an intolerable test. I hid in shack-like bars, in the company of old men, and made calls. Very soon my own Sahara crossing was in trouble: Akly in Agades was adamant that I could not cross to Algeria from northern Niger.

'Why not?'

'There is a war.'

'Seriously? Surely not . . .'

'It is too dangerous. The frontier is closed.'

'What is the danger?'

'The conflict between the army and the Touareg.'

'I have not heard anything about this.'

'No one has!' he cried, with a joyless laugh. 'This is the war that does not exist! But believe me, I am here in Agades, I have lived here for many years, and it is not safe to the north, and the border definitely not. It is not safe for you to go there and it is not safe for me to help you.'

The Touareg formally describe themselves as Imashaghen, 'the noble and the free'. They are also known as the Kel-Tamasheq, the people who speak Tamasheq. Their culture has been dated to the first century AD but it is almost certainly older. Traditionally there were four Touareg kingdoms, based in the four mountain ranges of the central Sahara: the Tassili, the Adrar, the Hoggar massif and the Air Mountains. The Touareg have always lived in a world of great spaces, relatively recently divided by frontiers: the entities of Niger, Mali, Algeria, Burkina Faso and Libya have all been imposed on their timeless homelands of sands and stone. Their first uprising, against the formation of the state of Mali, began in 1960 and was swiftly crushed. The second began in 1990: this time they fought the governments of Mali and Niger, in the name of autonomy, and again they lost, though peace agreements signed with both countries (with Mali in 1992 and Niger in 1995) called for the decentralisation of national power. Then, in February 2007, a group of Touareg, 'the Niger Movement for Justice', rose again. Niger declared a state of emergency in the north of the country: this conflict was now over a year old, and unresolved.

Stories from this unreported war were easy to find – I barely had to agitate the surface of conversations with barmen, waiters, idlers and drinkers for snatches to emerge. The rebels had mined a road near Niamey not long ago. They were at war with the government over

independence, and also for a share of the mines. The mines – didn't I
know about the mines? The uranium mines, of course! The ones the
French controlled, until the government had sold a slice of the contracts to the Americans. The same government, the same Americans,
the same French who did not care about anything in Niger as long as
the uranium kept coming, and the Touareg stayed quiet. The same
government who paid the army well, and the police, and cared
nothing for anyone else.

'There is nothing else in Niger,' said a barman, who spoke with an
extraordinary vehemence, a fury I had not seen anywhere else. 'The
tourists come for the cave paintings, they don't care; the Americans
and the French come for the uranium, they don't care; the government gets rich – they don't care.'

I watched the French Embassy staff at dinner. They were a credit
to their country, in a way: glossy men and glamorous young women
who would have fitted into the smartest Parisian restaurant, with one
or two older, tougher individuals among them. They ate under a hot
golden moon which hung in an indigo sky while a band sang and
played for them. They did not look around once: neither the
entertainment, the waiters, the life of the restaurant or the other
diners held any interest for them. In Niamey, it seemed, their business
was all the business there was. Later on, in a bar, I drank with soldiers
of a commando unit of the armed forces of Niger. They wore T-shirts
on which a grinning skull was wreathed with AK-47s. They were big,
rough men; a brusque, uncouth counterpoint to the civilised table of
diplomats.

I wanted to get out of Niamey as soon as possible. Perhaps I should
have taken the time to secure visas for Mauritania and Mali, and made
a wide semicircle overland, avoiding the trouble on the Algerian
frontier. But I was acutely conscious of days falling off my Algerian
visa, and desperate to get there. Algeria, I felt, held the key to my
journey, as much as to that of the swallows. So huge, so diverse, so
close to Europe – and yet so little visited, since the French left and the
country fell into civil war; a place of such dread reputation, and so
little known. I had wanted to go for years. The idea that the swallows,

our swallows, passed through it every spring and every autumn intrigued and beguiled me. I resolved to fly to Algiers, and greet the birds as they came in from the desert.

The flight went via Ougadougou and Casablanca, and left in the middle of the night. Waiting for it, in the half-lit airport bar, I saw the barman sit up suddenly, his posture stiffening with interest. An aeroplane was coming in.

'What is it?' I asked.

'I don't know,' he said, 'I have never seen that plane before.'

It was a jumbo, a Boeing 747, and it was all white. There were no markings on its tail and nothing obvious on its fuselage. As it taxied to a halt I pulled out the binoculars and studied it. Now I could see it quite clearly – it was a cargo carrier with an almost empty hold. Two pallets came out of it, and two pallets went on. On the side of the plane, in tiny lettering, I could just make out a registration number and an American flag.

'What do you think they're doing?' I asked a friend of the barman's, a woman in uniform who had also taken an interest.

'I don't know,' she said, then she laughed: 'Better not ask.'

It so often seemed to be the way in the Africas I passed through. If you did not know what *they* were doing, your gun-toting government, the foreign corporations, the British oil people, the French loggers, the American uranium hunters, the Chinese copper contractors, the mysterious Russian aircrews, why, then you had better not ask.

CHAPTER 8

The Walls of Algiers

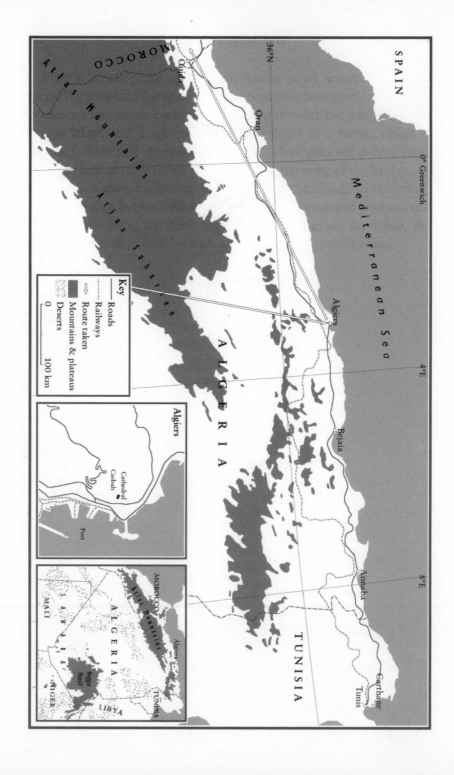

The Walls of Algiers

I did not know I was carrying contraband until they pulled me out of the line. The only European on the plane, nervous with antici-pation at what I would find, I was wondering why they X-ray the bags after you have landed when there was a commotion around the monitor, I was pointed out, uniforms closed in and they marched me into an office.

There was my bag, and my binoculars.

'Are these yours?'

'Yes.'

'What are you doing in Algeria?'

'Er, tourism. And bird-watching.'

Although the young woman operating the X-ray and the man who pulled me aside had been in uniform, everyone in the office was in plain clothes. They were all tall men, from young to middle-aged, and they all seemed to be smoking, except my interrogator, who had been halfway through a baguette sandwich. I pulled out my cigarettes and lit up too, the nicotine flooding me with artificial relaxation. The questioner studied my passport.

'Right!' he said. 'We will give you a receipt for your binoculars.'

'A receipt?'

'Yes. No binoculars in Algeria!'

'Why not?'

'Security,' he said, with a smile, and wrote. More questions followed and I volunteered more information than necessary, babbling about swallows and smoking too hard, nervously, but they did not seem to mind.

'Come and get your binoculars before you go,' said the man. 'Enjoy your visit to Algeria. *Bon courage!*'

'*Bon courage!*' was everyone's favourite valediction. The man at the Air Algeria desk, who revealed that a flight to Tamanrasset was cripplingly expensive and took three hours; the man who changed euros to dinars for me, and recommended a taxi, and the taxi driver himself.

'*Bon courage!*' they all said. It was a rather sweet expression and also unsettling. What was going to require so much courage?

The flight had been spectacular. Sometimes there were tiny bundles of light below us in the desert. The sky lightened as we approached the Atlas, their snows and cliffs of ice and stone protruding from rock-pools of cloud. To the migrating birds the lie of the land must swell and recede like a series of waves driving northwards. The Cape Flats rise up to the Drakenstein Mountains and the escarpment to the Great Karoo, which folds into the great veldt plains of the Free State, the Kalahari and the water-table of the Okavango, only to climb again to the high plateaux of Zambia, and sink into the green vastness of the Congo. The forest lifts to the mountains of Cameroon, drops into the undulations of Nigeria and the sandy seas of Sahel and Sahara, then peaks here, in the Atlas before subsiding once more into the fertile plain of the North African coast. There is nothing to compare with these mountains until the Pyrenees and the Alps, the ramparts of central Europe. As we flew along the coast of Algeria I could see, to the north, across the sea, a distant shore, and just above its far horizon, a little more solid than cloud, the snowy peaks of the Sierra Nevada – a first glimpse of Europe. On the other side, to the south, were mountains: some brown, some ochre and some very green.

The approach to Algiers is a little like the descent into Palermo: a city couched in a ring of mountains beside a wide bright sea. But where Palermo might have a ferry approaching and a few fishing boats

leaving its harbour, Algiers has dozens of tankers, container ships and
gas carriers coming, going or standing off at anchor, attendant. You
cannot miss Algeria's treasure. Even flying over at night, flares light
the desert; you certainly cannot miss it coming in by day as the road
from the airport takes you right past clusters of storage tanks like huge
steel pucks: Algeria has oil.

The light was the first thing that struck me, but then I had been
waiting for it. Like so many who are first taught to read French
through the strong simplicity of Albert Camus, Algeria to me was
Camus' country. Algeria's light, I expected, would be the pellucid,
piercing, lethal light that burns everything else out of the brain of
Meursault, the Outsider of Camus' *L'Etranger*, just before he
murders a man, 'the Arab', on a beach – and so it was.

We hummed along the motorway into thickening traffic and I
strained to take in all the brightness of the day. Like Sicily, Algeria is
a country in love with spring, and this bright blue day, spring had
come. Police with slung machine-guns eyed us as we crawled through
a checkpoint on the edge of the city.

'It's normal,' said the young taxi driver, and we drove into Algiers,
'*Alger la Blanche*', as the old postcards put it, as the French used to call
it, when it was the greatest treasure in their empire: 'Algiers the
White'. To see it is to think you understand, in the same way one
thinks one understands on first seeing Cape Town, why men kill in
and die for and cling so ferociously to particular pieces of the world:
Algiers is utterly beautiful.

It is built on the slopes of rumpled mountains where they tumble
down to the sea. Many of its roads are contour paths which weave in
and out of the ravines and re-entrants of these tumbledown hills;
streets seem to float like scarves in mid-air, wound around pine trees,
villas and apartment buildings, crossing bridges, narrowing and
doubling back like mountain tracks. There are gardens of palms and
sweet-scented plants whose names I did not know. There are
Ottoman palaces with bougainvillea pouring over their walls like
purple foam; lower, roads glide and swoop down boulevards, through
terraced white Art Deco fronts to the port's arcaded waterfront where

grand banks and great palazzo-like office buildings, once the long arms of colonial administration and commerce, gaze north, across the sea towards Marseilles. The port and the harbours are vibrant, working places. Where once the corsairs auctioned European slaves, ships now manoeuvre, trains pull in and out of the station and I watched four cats confused over who had what right to which of five fish-heads.

Just to the west of the centre, climbing away from the sea up its own hill is the Kasbah, Turkish and pre-Turkish, an ancient, riddled and riddling world-heritage warren, perhaps best known to the world now as the battleground of the Battle of Algiers.

At the gateway to the hotel the taxi stopped while a man in uniform probed underneath it with a mirror, then poked around in the boot with a chemical bomb-sniffer. By the time I had reached my room, opened the windows and stepped on to the little balcony, into a view full of sea and white roofs and a blue sky spinning with swifts, I was smitten. I had only five days in Algeria. I would not spend them waiting for buses or watching the hundreds of miles of country pass by a window. The swallows would have to come to me; I resolved to give all the time to Algiers.

The commitment to smoking is remarkable: should you want to light up three times between the lift and the reception desk, or extinguish five cigarettes between the concierge and the main door, there was an ashtray on hand for every match and flick of ash. Algerians are champion smokers. Everyone over the age of seven has lived through worlds of death unimaginable to us. Their story renders ridiculous our world of nebulous health-scares.

'This race, wholly cast into its present, lives without myths, without solace', Camus wrote. 'Everything that is done here shows a horror of stability and a disregard for the future.'

This essay, 'Summer in Algiers', was written in 1936. The race he refers to is his own: the mixture of French, Spanish and Italian '*colons*' or '*pieds noirs*' who had colonised the country since the French conquest of 1830. That race has vanished now, scattered to disappearance. Though it refers to a lost world, 'Summer in Algiers' is still full of contemporary truth: only the occasional line has been rendered anachronistic in the most terrible way, as if it took an apocalypse to eclipse Camus' observation. All its ten short pages are shot through with something dreadfully beyond irony: with prophesy.

> This country has no lessons to teach. It is completely accessible to the eyes, and you know it the moment you enjoy it. Its pleasures are without remedy and its joys without hope . . . In the Algerian summer I learn that one thing only is more tragic than suffering, and that is the life of a happy man.

The hotel gardens were full of exotic plants, and gardeners, and no guests. Just beyond the gates was a garage and beyond that a road leading up to the British Embassy. Though I called them, left messages and knocked on their door I never roused any response from that embassy: I had hoped someone there might give me a briefing on Algiers, perhaps furnish me with Algerian contacts. Two policemen on duty nearby said they saw very little of the diplomats. It was not surprising. In the autumn of 1993 war was declared on foreigners: by the summer of the following year over fifty had been killed and the French ambassador had ordered the last 2,000 French expatriates to leave. Every life is of equal value, and many of those killed died horribly, but still, since 1962, when the Algerian war of independence ended with the mass exodus of the *pieds noirs*, fifty foreign deaths are a tiny speck in the maelstrom of Algerian blood which engulfed the country before and since the French left.

The two police officers were dressed as if for a musical featuring Italian policemen of the 1920s, in bright sky-blue uniforms adorned with sparkling white cuffs, belts and straps, and large caps. They would have looked comical, toy-like, had it not been for their automatic weapons: every Algerian policeman seems to have a well-oiled assault rifle within easy reach. The guns were striking because they seemed so well looked after, their wooden stocks and grips smooth and shiny with long use. But at least you can see their owners' faces. In 1993 the state created a force of 15,000 paramilitary police who came to be known as the Ninjas, because of their black balaclavas. Their job was to kill and torture: to kill the Islamist killers and torturers, who themselves killed indiscriminately, and to torture those suspected of harbouring or sympathising with the killers. Men like the Ninjas – who might have been Ninjas, how could one know? – faceless, unaccountable men also murdered civilians who might sympathise with the FIS, the Islamic party which won the 1992 elections, triggering an army *coup d'état*. By the late 1990s anyone could be killed, in fact: slaughter served to weaken the state the Islamists hated; slaughter also served to cull those who might vote for Islamic parties. The army, the terrorists, the intelligence services and the paramilitaries, all stoked the horror, and fed on it.

Death became the *raison d'être* of the GIA, an acronym initially deployed to describe a loose network of Islamist killers, but one which came to be questioned by investigative journalists: the massacres ascribed to the GIA often seemed to serve the interests of the army generals and the mafia of shadows who comprise '*le Pouvoir*' – 'the Power' – the vague and loathsome force which has squatted on or near the levers of Algerian power since 1962: the men with secret bank accounts stuffed with the proceeds of Algerian oil.

Algiers was bright and beautiful, that first day, but unsmiling. The killing has abated: the country is led by President Bouteflika, a prized ally of the West since 2001 and the coming of 'the War on Terror'. Occasional bomb attacks, roughly one spree a year, are now ascribed to 'Al Quaeda in the Islamic Maghreb' but few trust this provenance. They keep the population docile, it is pointed out, and *le Pouvoir* is still in power; and behind Bouteflika, and who knows what else, is the DRS, the Département de Renseignement et de Sécurité, a secret service founded in 1990 principally to protect and serve the generals at the heart of the *Pouvoir*, which specialises in torture, assassination and kidnap. The lessons Algeria has to teach are almost too awful to study, which perhaps explains Europe's attitude towards it. We have heard and forgotten reports of coups, massacres, earthquake and

bombs – otherwise, apart from the oil workers, flown in and out of
fortified camps in the desert, and the never-mentioned US base at
Tamanrasset, the West has largely ignored it.

'Don't you find it oppressive?' I had asked the young taxi driver.
'There are so many policemen!'

'No – why?' he said. 'They are here for our security . . .'

Sometimes it seemed half the male population of Algiers was
employed to keep an eye on the other half.

Swinging down a long sloping street towards the centre of town,
marvelling at swooning-sweet gardens and tall palms below the road,
I happened to glance up at a passing bus. Every single passenger on it
was staring at me. I raised this with a friend I made at an internet café.

'Am I the only European in the whole city?'

'No,' he said, 'I saw another one yesterday.'

'Don't Europeans come here, then?'

'Oh yes – Italians, and the French, in the summer. You are a little
early, that is all. *Bon courage!*'

At the bottom of the road, where the streets opened into squares and
the boulevards descended to the sea, a fight broke out. My only other
experience of the testosterone overspill of young North African men
being in Morocco, I was shocked by the speed and fury with which this
battle ignited. Moroccans seem to love the idea of fighting but they do
not actually fight much – two youths dancing around each other making
a lot of noise and show is not an uncommon sight, nor particularly
disturbing. But these two Algerians were trying to kill each other. A
group of men formed a ring around them but made no effort to separate
them. Then police arrived, in force. It was strange to see such rage and
hatred in the middle of a sunny spring afternoon. Immediately, up and
down the street, older men went into gossip huddles.

When they were young these men were known as *hittistes*, from the
Arabic *heta*, meaning wall, a reference to where they spent their lives,
hanging about on corners, leaning on the walls of their neighbour-
hoods. Many have spent the greater part of their lives out of work.
After independence, following a coup in which Houari Boumédienne
deposed Ahmed Ben Bella, figurehead of the liberation struggle, a

flush of rising oil prices buoyed Algeria through Soviet-inspired industrial and agrarian semi-revolutions, into an urbanised modernity. In time this period, the 1970s, would come to be seen as a proud time, even something of a golden age. But by the mid-1980s, with Boumédienne dead and his successor, Chadli Bendjedid, ineffective in the face of corruption, a permanent gap opened between the rulers and the ruled. The population doubled to 23 million as oil prices fell. Along with mass unemployment, food queues and water shortages came to Algiers. Exit visas were abolished. The *hittistes* fulminated against the *chi-chis*: those connected to the elite. Hatred of bureaucrats, capitalists and politicians became street currency, expressed in bitter jokes and salved with zombretto, a cocktail of industrial alcohol and lemonade.

The people on the streets were a mixture of the secular, the religious and the westernised, at least by their dress. As many women wore jeans as headscarves. Most of the men seemed to wear leather jackets, which made them look like secret policemen. I attracted many stares, but nothing like hostility. In a bookshop, in a café, idling along above the harbour, I fell easily into half a dozen conversations: each time my interlocutor was anxious, almost desperate, to talk and be helpful. I caused a small pandemonium in the bookshop, expressing an interest in one volume which the assistant was sure he had in a better, unscuffed cover. His colleague was summoned, other customers joined in, books fell from shelves, voices were raised in exasperation: Algerians, I concluded, over the following days, give the impression of anxiety because they are perfectionists. If a man can see you doing something which he can also see could be done better he will not be able to resist telling you about it, and the longer you desist from following his advice, or the less perfectly you follow it, the louder he will inadvertently shout in attempting to assist you. The effect of this is that every now and then streets, cafés, corners or squares resound with raised voices, but what sounds like a widespread and angry dissent is in fact perfectly normal exchange.

*

The sweep of Algiers leads you to the sea: the gradients flatten along
the front and then, like the gentle swell of a wave, carry you up to the
Kasbah. There is a small square just before the main road into it,
where one swallow flew round and round a cluster of stalls in the
middle. She was a female, alone, and there seemed something hectic,
something lost about her, like a bat caught out by the daylight. I
stopped for coffee. The stall was busy; two or three serving and an old
man in charge. One imagined he had donned his old blue coat, such as
grocers used to wear, every working day for years. His pate was dark
brown and his eyes were black, with dark semicircles below them.

'French?' he demanded.

'No – *Gallois!*'

'*Gallois* . . . Tourist?'

'Yes!'

'Why have you come here?'

'Just to see, to find out, I have read much about Algeria, and I have
always wanted to see it.'

'Ah, it's good, it's good. What would you like?'

'Coffee, please.'

'Give me a coffee here! So what do you think of Algiers?'

'*Formidable!* It's beautiful . . .'

'Yes, yes, beautiful . . .'

'Have you lived here long?'

'All my life – ha! A long time.'

'So you have seen a lot.'

'Ha! A lot! Yes, everything. I live in the Kasbah. Have you seen the
Kasbah?'

'No, I am going there next. So – you remember the war?'

'But of course. It was here. It was on this street. People died there,
and there, shot down by the paratroopers. Did you see on the corner?
There is a plaque. And another across the street. I remember the boys
who died there.'

'It must have been terrible . . .'

'Terrible. They did us wrong.'

'The soldiers?'

'The French did us very wrong,' he said, with a sorrow, not an anger, as though he was not describing an assault so much as a betrayal.

'How did you – live?'

'In the walls. We lived in the walls. Do you understand? We lived in the walls to survive.'

He would not let me pay for the coffee and when we shook hands he held mine for a moment. 'You must go into the Kasbah and see. It is good that you are here. *Bon courage!*'

'*Merci, monsieur. Et à vous aussi.*'

The war began in 1954. By 1962, when it ended, more than a million Algerians had died, six French governments had been brought down by it, de Gaulle had been returned to power to face it, and his solution to it, the eventual return of Algeria to the Algerians, had come extraordinarily – amazingly, it seems now – close to bringing a civil war to French soil, as entire regiments of the French army in Algeria mutinied against what they saw as a surrender which made meaningless all the terrible things they had done, and betrayed all those they had lost.

The war in the Kasbah – called the Battle of Algiers – can be seen as the paradigmatic heart of the conflict. Up and down the tiny streets, which seem to wind, rise and fall in three dimensions through the packed, stacked, clay-coloured nest of this ancient town within a town, a maze of tiny doors, peeping windows and secret staircases, French paratroopers and the Front de Libération National engaged in a pitiless struggle. By all the rules of insurgency, in the midst of a hostile population, against an enemy that was born to the battlefield, the French, not long since defeated in Vietnam, should have lost the Battle of Algiers. But they did not, and this has come to be seen by the mighty but mired occupying armies of our own time as the terrible lesson of Algeria: the model of how guerrillas can be beaten on their own ground. The formula is as simple as the great negation of evil itself. You win by torture and summary execution.

A paratrooper, Pierre Leuillette, recorded his experiences in an interrogation centre, in a disused sweet factory.

All day, through the floorboards, we heard their hoarse cries, like those of animals being slowly put to death. Sometimes I think I still hear them . . . All these men disappeared . . .

The numbers of the disappeared are still disputed. From the summer of 1956 when FLN fighters gunned down forty-nine civilians over three days in June, to the end of March 1957, around 3,000 men and women were seized and killed, many having first been tortured. Had the FLN won the battle it is likely the war would have ended that year: instead it continued for another five.

Before she was killed in 2006 the Russian journalist Anna Politkovskaya wrote about what she called the Chechenisation of Russian society. The horrors Russia visited on the little republic have warped and perverted Russia's own soul, she said: in killing and torturing, beyond any law or justification, Russia has carved a capacity for numbness at horror and debasement into its national psyche. Something similar seems to have begun to happen to France: the difference was that those who spoke and wrote about it did not themselves become state targets. French soldiers, writers and intellectuals drew parallels between French methods in Algeria and those of the Gestapo. Something in the heart of morale, any army's ultimate weapon, died in the torture chambers, in the shallow graves, in the body-dumpings at sea, in 'the work in the woods', as the soldiers called the business of secret execution. In crushing the Kasbah, all sides later agreed, France won the Battle of Algiers and lost the war for Algeria.

The songs of caged goldfinches trickle sweetly through the alleys of the Kasbah. Then the rain comes, hard cold curtains sweeping in from the west, and wind: climbing steep, cracked steps I turn around and the sea seems vertiginously close, as if a backward slip would tumble you to the streaming waves. A gas tanker rides the swells, no sign of life on her, as if the crew have deserted. I duck from shelter to shelter, hiding from the rain in cafés and, once, a workshop, where an old man fondles the ears of a tiny dog. In each place I meet the same questions, the same bemusement, which each time turns to something like friendship. In three stops in the Kasbah I am warned three times to be careful in the Kasbah, as if it is a den of thieves. I see no thieves. Veiled women hurrying out of the rain, children being called in, a shouting group of sixth formers coming home from school. I will not be here long enough to gain an invitation: I can only climb steps, wander alleys, stop, start and peer and try not to be seen to pry. I keep seeing the same wry smile on the faces of the men I speak to, but I cannot quite ascribe it. Am I the first tourist of the year, a premature sign – or am I an irony?

In my new book, *Ten Walks in Algiers*, a French resident of the 1920s mentions swallows. 'They love the old Turkish walls of the Kasbah', he says. I climb up when the clouds begin to break, and a washed white sun shines on gutters running black, into a smell of pines, and suddenly a shadow flies across the street at my feet and I know before my gaze finds them that they are here. Two, five, seven at least – and there are the old Turkish walls and there they are – and they still love the walls of the Kasbah! The light makes their breasts and backs shine brilliantly as they throw themselves up and over the street in that way they have, as if the downbeats of their wings render them weightless, to catch and redirect themselves forward on the upswing.

In the next three days I come to know their habits. A small population of 'our' swallows, Barn Swallows, breed on the North African coast. Among those I watch in Algiers, some appear to be prospecting for nest sites: there is no other explanation for their attachment to certain parts of the city. In the coffee shop square the lone female now has companions; higher up three or four birds hunt a particular switchback street near the Kasbah walls, and then one morning in a sun bright as knives, which would have been hot were it not for the wind, I find others between the city and the shore. These do not seem to be prospecting. They are on their way north, I guess, and feeding up for the flight across the sea.

The long white arcade pullulates with people and the traffic is thick

on the road beside it, overlooking the port, and the square around the
corner is packed. I am coming to terms with the city's strange rhythm.
Friday is mosque day so everyone is off, which means Thursday night
is a bit like Friday night in the West. Saturday is off and the week
starts again on Sunday, unless you are on a European time, which
much of the city is, in which case you do not go to work until Monday.
If I lived here, I daydream, I would only work Monday, Tuesday and
Wednesday: the rest I would give to writing, reading, walking and
making love.

> In Algiers whoever is young and alive finds sanctuary and occasion for
> triumphs everywhere: in the bay, the sun, the red and white games on
> the seaward terraces, the flowers and the sports stadiums, the cool-
> legged girls.

The intense physical beauty of Algiers, which Camus caught
perfectly, still sings through the movements of people: in the way they
carry themselves, the way they look at you and each other – the
moment-to-moment seems very vivid here; a migrant such as I can
only speculate but it is as though years of unemployment and war and
horror have culled people's faith from the future and focused their
concern on the present.

Or perhaps it is only that this is such a sparkling late morning, and
that coffee tastes particularly rich in the mouth and cigarette smoke
particularly satisfying; that the ship docking at the quay is particularly
huge and its manoeuvres particularly delicate. The rushes of sea wind
seem to fan the sunlight and brighten the glare. And the swallows –
were they human you would have said they were high on it all: dozens,
darting in and out of the colonnades, flinging themselves into the
traffic, dodging buses and scooters with equal unconcern, sling-
shooting themselves in the strongest wind at the corner, whipping
back 90 feet only to skew sideways and back under the arcade. All the
thousands of miles they have flown seem to have distilled into
confidence: there is an exultation in their speed and fitness, as though
they are playing in the waves of the air, as though they have mastered

all they have crossed: the savannahs and hills, the storms of the forests, the deserts, the mountains – as though they have conquered Africa. I stood at the corner grinning like a fool and watched them dive into the square. Suddenly they were down, no more than 3 feet from the floor, cutting through the crowd at leg-level. Either they played, flew for fun or were showing off to each other – this could be courtship or celebration. They do indeed find sanctuary and occasion for triumphs in every square inch of the air.

I begin to walk along the sea front, determined to follow it westwards until my legs give out. Many men, women and children appear to have had the same idea, or, at least, to have been drawn to the sea. Perhaps it is the weather. Couples sit on benches, not quite canoodling. Families straggle, women lean on balustrades, men fish: it is as though in crossing the road between the city and the sea one steps into a municipal holiday. I walk into one of the many bands of boys playing football on one of the terraces.

'Hey!' one of them says.

'Hey yourself!' I return, and am instantly surrounded by children.

'Who are you?'

I tell them my name.

'Where are you from?'

'Wales.'

'What is Wales? Where is it?'

I explain. They look dubious. Then one boy brightens suddenly, with an excited smile and a cry: 'Manchester United – Ryan Giggs!'

'Yes!' I say, equally thrilled. 'Ryan Giggs!'

It is not the first time someone has made this connection between the strange and the celebrated, the obscure and the familiar. It has been first a curiosity to me, then a suspicion, then a certainty that throughout Africa it is an absolute fact that the Land of Song, the armoury and workshop of the Industrial Revolution, the country of Snowdon, Gareth Edwards, the Arms Park (as was), the Great Thomases – R. S. and Dylan – the Dragon and Charlotte Church exists only in answer to a question about English football. This is: if England are so rubbish at football, especially compared to the

wonders of its Premier League, why on earth do they not put Ryan
Giggs on the left wing?

Ah – some Man. U. fanatic must have answered a thousand times,
in a thousand football conversations, in a thousand distant villages and
towns – because Ryan Giggs is . . .

'Do you play football?' another boy demands.

'Ah. Not really, I'm afraid – I'm a bit rubbish. I play rugby though
– do you like rugby? It's the national sport of Wales.'

They were not in the least interested in this.

'Oh, come on! Play with us!'

I do my best but become rapidly frustrated at being effortlessly
dispossessed by Algerian nine-year-olds, laughing cynically as they do
so and calling me 'Reean Geeggs'. In the end I pick up the ball and
charge an imaginary try-line.

They let me go after that, to walk on and ponder that if the
homeland you love has to exist merely as a trivia point subordinate to
a sportsman it is a great and fitting consolation that he should be the
Zidane of the touchline, the Cantona of the cross-in, that human burst
of speed and grace – Giggsy.

Next I meet the friendliest secret policeman in Algeria. I am
sceptical at first. He hails me from the other side of the road. I greet
him in return but he shouts again, dodges through the traffic and
accosts me by the sea wall. He wears jeans and a pale jacket, a wide
smile on his thin face.

'Don't worry!' he cries. 'It is OK, I am a security man – I am here
for your security.'

This phrase is beginning to irk me.

'My security is fine, thank you,' I say, and make as if to move off.

'Do you know people used to live down there?' He points to a
cluster of shacks and houses built below the wall, just above the sea,
on collapsing pillars.

'And there are some still there, but there used to be lots – fishermen
– do you see?'

'I do, yes,' I say, and surrender, gracelessly, to a conversation he
seems over-keen to have.

'You are not the only tourist, no, no!' he says, when I ask him if I am some sort of rarity. 'They come in the summer, now, Italians and French.'

'The French? Isn't it strange for them?'

'Strange, no – why?'

'Because of the war.'

'No, no. This is all in the past! We do not live there now. Every nation does bad things, doesn't it? Every people in history, but we have to forgive. We do not hold anything against the French. I like the French! Do you know, there were French people here during the war who were kind to us, who helped us? Who hid people the soldiers were looking for? Well, there were. There is an old French couple who are my friends who have always lived here. They love it here, they say they never want to leave. And they know they are welcome here – and you are welcome here! We are very happy to see you in Algiers.'

In the hotel that night I ask a man in a big leather jacket for a light. Ali is smoking a cigar. We talk, and he looks distressed that anyone should contemplate eating in the dining room when there are better places in town. We are soon in a taxi, whose driver Ali knows, and then in a dark street near the port where Ali rings a bell. A door opens, cautiously at first, then is flung wide when he is recognised. Inside, we pass through light and warmth and a press of men drinking beer to a staircase which leads up to a room in which men and women crowd around a U-shaped counter, which surrounds a smoking grill. We eat succulent lamb chops and drink lager as Ali talks of the wonders of Algeria I will not have time to see. The Kabyle, he says, in the north-west, is like Paradise. Fruit trees, orchards, valleys of flowers, sharp-crested mountains, virgin coves fringing a crystalline sea: 'You must go to the Kabyle!' he cries. 'But you must go with me or someone who knows it, a good driver, because there is still trouble there. They hate the *Pouvoir*.'

And Oran, he said, you must go to Oran, and beyond, round and down into the desert, where his family lived, I would always be welcome there. Ali was a businessman; I was never quite clear what his business was, exactly, except that it took him to France, to Lyons, often, and to Marseilles, which he loved because it was like Algiers, and to our hotel, where he stayed when he was in town.

'This man', he introduced us, 'is the patron of this restaurant. His father ran it before him. This place has survived everything – this is the true Algiers!'

There was an air of freedom and relish laced into the meaty smelling smoke from the grill. Women swigged bottles of beer, men laughed and flirted; we all seemed to smoke, eat and drink at the same time.

'In Algiers no one says "go for a swim" but rather "indulge in a swim", Camus writes. 'The implications are clear.'

I thought of places where I wanted to live, of Tsumeb, of the valley between Zambia, Mozambique and Malawi, and of Makoua, and I thought of places where I have lived, like London, Palermo, Grenoble and West Wales, and it seemed to me that I could find something of all of them in Algeria, and I did not want to leave. The next morning my visa expired.

The border between Algeria and Morocco is closed, the conse-quence of an old dispute about Algeria's support for the people of the western Sahara whose land has been annexed by Morocco. One can only marvel at this mighty handicap to the entire region's develop-ment: Algerian resources and Moroccan connections to Europe and the West would be a formidable combination. The only way to the Straits of Gibraltar, the great crossing point for people and birds, is via air. My flight went to Casablanca. At the airport they returned my binoculars. I will come back, I swore, I will come back by sea, accompanied, when I have found her, by the woman I will marry, and I will give something of my time to Algeria.

CHAPTER 9

Moroccan Tricks

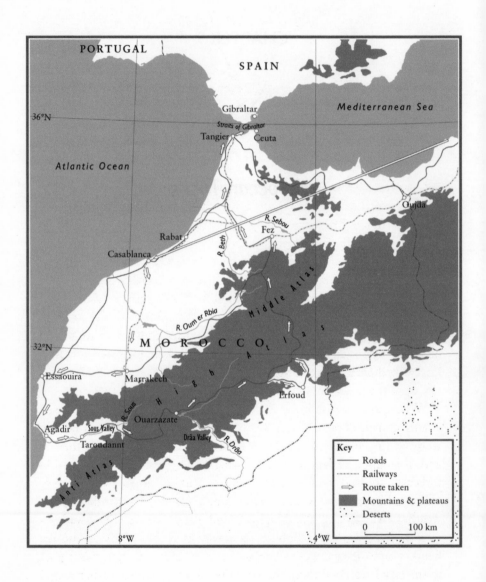

PORTUGAL

SPAIN

Mediterranean Sea

Gibraltar

36°N

Straits of Gibraltar

Tangier   Ceuta

Atlantic Ocean

Oujda

R. Sebou

Rabat   Fez

R. Beth

Casablanca

R. Oum er Rbia   Middle Atlas

M O R O C C O   A t l a s

32°N

Essaouira   Marrakech

H i g h   A t l a s   Erfoud

R. Sous

Agadir   Sous Valley   Ouarzazate

Taroudannt   Drâa Valley   R. Drâa

A n t i   A t l a s

Key	
——	Roads
----	Railways
⇨	Route taken
▧	Mountains & plateaus
∴	Deserts

0   100 km

8°W   4°W

Moroccan Tricks

It is late March, springtime in the Mediterranean. Fleeing the horrors of the European winter, which this year has been cold and brutal, refugees from the higher latitudes converge on the rim of North Africa. Exhausted by their jobs, drained by the strain of months of rain, darkness and prolonged exposure to newspapers, the travellers, tourists, families, lovers and bargain hunters come in their thousands, heading for the Jerusalem of the Leisure Age, Marrakech, the city under the High Atlas. For the fortunate majority, a swift cheap flight brings relief, and freedom. But for those who are not so lucky, or more curious, or who simply hold their lives and fortunes less dear, there is a notorious stopping point on the way to their year's first sun: the greatest port of the eastern Atlantic – Casablanca!

'Where are the swallows?' I kept being asked, in text messages and emails, during those days in Morocco. Friends and family in Britain were waiting for them, straining to see them, worried about them – and I laughed. I knew exactly where they were. The television forecasts showed rain and snow over Britain, storm systems over western Europe, and a clear break, like a bubble of warmth and fair weather over Spain, the Maghreb and West Africa. The swallows were moving with this bubble, as surely as if guided by an unseen hand. The vanguard were all around me; on the road from the airport to the city I watched them hunting the edges of fields, cutting in and out among tall trees. In rainy weather when there are not many flying insects, the birds sometimes brush trees and bushes to dislodge prey.

I had been to Casablanca before. The first time I arrived, the taxi drivers at the airport refused to take me into town, saying it was too expensive – why not take the bus? But things had changed; now it was all hard bargaining and my driver was moody, unsatisfied at the price we had agreed. Casablanca is the most European of Moroccan cities. The traffic and the pace of life have a European bustle about them. It is as though, so close to the bright lights and bank accounts of Europe, something of Africa's ease and philosophy have been burned away, as though our wealth acts like a fire, searing away the human in favour of the economic. At the same time, Casablanca is and always has been a trading town, a centre of business and commerce, and it is home, I knew from experience, to the champion hustlers of all Morocco. The second time I came here I ended up financing an impromptu holiday for two girls, one brother, two children and a cook. This time, having come so far, I was confident that I could take care of myself. I was, in fact, over-confident.

Edith Piaf used to stay in the hotel I chose: perhaps I was lavish. Certainly I went out for lunch, without change. The beautiful blue 200 dirham note is a flag with 'eat me!' all but written on it. I tried, at a very pleasant little hole in the wall full of students where a charming kebab-seller prepared a delicious kebab and a helpful young man was doing bits and bobs: tearing up paper for napkins, etc. The bill was 15 dirhams (including a drink and all the trimmings); I gave the young man the note and never saw either again.

I was furious and the kebab-seller most distressed. He had no idea who the boy was, he protested. I waited, hunted about a bit, went around the block, stalked the hole in the wall like a tiger creeping up on a lamb . . . It was barely therapy. And there were still the 15 dirhams. I said I would pay tomorrow, after I had caught the bastard, and strangled him. Smiling sadly, the kebab-seller said I could have it on the house.

It would have been a good moment to go down to the port, to kick through the sardine and diesel juice on the quays and admire the crews lounging about in the sun on their wooden fishing smacks.

Instead, bloody-minded, I decided to go back to Edith's, unpeel another note from the dwindling roll, and have the afternoon I had planned: a walk, a drink somewhere, maybe an adventure, who knows, a fish supper: it was Casablanca, after all.

'Be careful!' said a voice, as I stepped into the road.

'What?'

'I said be careful,' he said, reasonably, as we crossed together. It was barely a street, more a break between two of the new town's shaded colonnades. He was a little taller than me, with a handsome nose, eyes like bright wet ink and a slightly threadbare tweed jacket.

'The road.'

'Thanks,' I said witheringly. 'Like a baby?'

'*Pardon?*'

'*Merci bien*, but I don't need help crossing the road.'

'Excuse me. It's just that round here they drive very badly. The roads here are very dangerous.'

'I know. I have been to Casablanca before.'

The first and so far only car chase I have been involved in had begun less than quarter of a mile from where we stood, but though there is a certain flamboyance about Casablanca's highways, they cannot really compare with anything Nigeria has to offer. Those motorbike rides in Calabar, for example. I must have been crazy.

'Oh, you have visited before?'

'Yes, twice.'

We were walking along, chatting easily. He had a kind of rolling gait.

'I thought you were new,' he said, giving me a sidelong look.

'Do you live here then?'

'I am from here but I do not live here. I am a sailor.'

'Oh really? What sort of ship?'

'Cargo.'

'Wow! A proper ship!'

He shrugged. 'I am a mechanical engineer – an electrician.'

Mustapha was born in Casablanca. When he was young he made friends with an older man, an engineer, who took him on as an apprentice. He had lived in Holland for a while, on a barge, until his best friend's wife made a pass at him. Mustapha rejected her, but, scorned and wrathful, she said something to Mustapha's best friend, who threw Mustapha out.

'And I did not say anything. Nothing! I thought this is your wife, I am your friend, but you have chosen to believe her. You do not think I am an honourable man, but I am. So I left and did not see him again for many years. Then one day I met him again. He had come back to Morocco. He begged me to forgive him. I said I did. He said he had come home one day to the boat and found his wife in bed with his brother.'

'Do you still see him?'

'No, I have not seen him for years. But that is life . . .'

Mustapha's ship was on a run down the West African coast; the next stop, at the weekend, would be Agadir, where Mustapha would see his wife and children again. We sat in the back of a café, drinking coffee and smoking, waiting for Aziz. The plan was that when Aziz showed up Mustapha would replenish his supply of hashish and then if I felt like it I was welcome to go with him to visit his boat. I could not wait. A proper ship in the port of Casablanca and a good conversation – sea stories no less! – and, if I felt like it, a shot of decent Moroccan hash: what pleasures.

We discussed dope. Mustapha had smoked it all his life. He never went anywhere without it, he said.

'Look,' he had said, as we swung along the road on the way to the café, withdrawing his right hand from his pocket. Stuck to the tip of his index finger was a tiny blip of dark resin.

It did not look appetising.

'But don't you worry about the damage that it can do?'

'If you smoke it properly it is quite safe,' he said. 'You need the very best hashish, and you only smoke two joints a day. One after you have eaten, and one before you go to bed. And Aziz only sells the best. He is a big dealer, but he sells me a little because we are friends. We always meet in Casa. I am lucky to know him . . .'

I was not really very interested; every stoner will tell you his hash is the best, and his dealer is good, and he is lucky to know him, but I was polite. Oh really, jolly good, lucky you . . .

'If you like you can buy some too.'

'Oh, you're very kind, but I don't think I will. Perhaps you will let me try a little of yours . . . ?'

'Of course! It is up to you.'

We talked about the shipping business. Mustapha's ship was a dry bulk carrier. I was very interested in the trade routes – cereals coming over the Atlantic from North America, being unloaded at various ports down towards the Gulf of Guinea, and being reloaded with . . .

Aziz appeared. Slightly overweight, with a yellow-olive skin, glasses and better clothes than Mustapha's, he was sweating slightly, having hurried. Who would be a dealer? Poor fellow, really. It had obviously made him good money, but what a life. Always hurrying to a rendezvous. Being so careful with your phone. Having to make instant decisions about who to trust, and how far you could trust the trusted not to betray a confidence to someone who might talk carelessly; knowing that half the demands on your time and who knows how much of your friendships are based on desire for what you peddle, regardless of anything you are.

We were rapid and scrupulous in putting each other at ease. We shook hands. Aziz, laughing slightly shamefacedly, apologised for his breathlessness. I offered him a drink, he gratefully accepted a mint tea, and he told me as much as he could about his life. He worked from his car. He never used a phone. He only dealt to a small circle of people he knew very well, and he did not get involved in this sort of thing, small deals, because it was not worth it. But then he and Mustapha went back a long way. The two old friends were very pleased to see each other, and I leaned out of their conversation, politely, as they worked out their exchange.

'How much do you want?' Aziz asked, after a minute.

'Oh, nothing. Don't worry about me. Thank you, though! It's a kind offer . . .'

'The thing is,' Mustapha said, reluctantly, 'I am going to buy 800 dirhams' worth. It is pollen, the very best there is, and it really is expensive, but Aziz gives me a very good deal. Do you want 400 dirhams?'

'No, really, you are very kind, but that's too much.'

I was curious. Pollen. I had heard about it, I thought. Such a lovely word. And what would the effect be? A sort of hashish equivalent of Bollinger, I imagined.

'A little, 100 dirhams would be great – but I don't want to carry a lot with me.'

The package passed from Aziz to Mustapha very quickly. It was about the size of a box of cigarettes, but more bulked out, wrapped tight in brown sticky tape.

'May I see?'

We were sitting at the back of the café and no one was paying us any attention, except the waiter, and Aziz was keeping an eye on him. We had all done this before. The package was in my hands without anyone suspecting anything.

It was soft, under pressure, then hard. I sniffed it, but the wrapping was extremely tight. I gave it back to Mustapha.

'The thing is,' Mustapha said, and I could see it pained him, 'Aziz . . .'

'I can't divide this any smaller,' said Aziz, surreptitiously checking the time on his phone.

'If you took half . . .'

'OK,' I said, trying not to sound weary, 'I'll take half.'

We worked it out very quickly after that.

It was not complicated. There was a guy who owed Mustapha money. We had to go and get it off him. Luckily I had some euros: Mustapha could get a very good exchange rate on them. If I paid the whole 800 dirhams now, and let Mustapha change some euros for me, Aziz could be on his way, Mustapha could give me the 400 dirhams, and, thanks to the exchange rate, we would both make a little extra on the euros, reducing the price we paid for the pollen. We headed off as quickly as possible, swinging by my hotel to pick up the euros, and some more dirhams (I had only come out with 400) and nipped into another café to do the exchange. Mustapha wrote down his name,

Mustapha Lotfi, and the quay number of the ship – I was welcome to come with him now, of course, but I was quite happy to go down later. I was tired, the afternoon was hot, and it was time for a Casablanca siesta; best taken, I thought, on my bed, under my open shutters, admiring the blue picture postcard shadow thrown by a tall palm onto the honey-yellow wall opposite, with a nice – indeed, superlative – spliff, and ruminations on Edith Piaf. Aziz, sweetly, gave me a token of appreciation.

'This is a present for you, you understand,' he said. A corner of delicious-smelling hash, light butter-gold with a darker crust.

'Thank you, Aziz! That's really, really kind. Would you like another coffee?'

'No, thank you, I really must go – I am late.'

'Of course. And how would I find you, if I am ever in Casa again? Phone?'

'Ah, I do not give out my phone – the police,' he said, pointing to the sky.

'Ah, yes. Sorry . . .'

'Through Mustapha,' he said.

'Through me,' Mustapha confirmed.

It would be easier, we decided, if Mustapha came to my hotel at seven. There was no hurry. I had the pollen, after all. We parted in a flurry as Casablanca's siesta hour began to melt into a threshing, gilded coil of rush-hour traffic. There was a bus pulling into a stop, a little taxi trying to get out of the stream of vehicles to pick up Aziz; I was going one way and Mustapha another. It was a job just to shake hands properly and not be run over.

'Look,' he said suddenly, quietly.

I looked. There was his finger again, with the tiny blip of hash on the end of it.

'That's class, eh?'

I was not sure when the dream began to fade, but now I see it again. I see myself hurrying back to the hotel, trying not to hurry. And not opening the package, because it would not be right: Mustapha ought to do the division. And starting to smile, and being filled with comical dread at the same time. And being tossed and tumbled in a wild double current: calm certainty in half my brain; rampaging incredulity in the other half. And then knowing, almost in the instant the door to my room shut behind me, knowing absolutely, and howling. The expletives would not form properly because my smile kept getting in the way. The fury would not ignite fully, because I could not – it was more physical than mental – bring myself to count right up to the number of dirhams and euros, converted into pounds, that I had spent and lent. Every time I came close to the total something would snap and I would find myself trying to force a fist into my mouth.

And then there was the package. I could decide it was pollen, break in and smoke a bit, just to prove it. I could wait for Mustapha. He would not come at seven, but of course I would wait for him anyway, on general principle.

'But it's beautiful!' I kept crying out. 'They were just – beautiful! Con artists? Con maestros.'

In the end I unwrapped a few turns of brown sticky tape, sniffed it and sneezed: snuff.

That was not quite the end of Casablanca. My bag now contained an additional £300 worth of snuff, and very little money. Thanks to Mustapha and Aziz, I entered my room an idiot, but I left it resolved to become a Berber. In the mighty market that is Morocco, there is no higher compliment a Moroccan can pay your bargaining skills than to call you a Berber. I still went out that night (as far as the disco in the basement of the hotel) but I drank beer from a corner shop, and the next day I took the bottles back and claimed a couple of dirhams for the recycling. I then walked to the railway station, Casa Voyageurs, via the kebab-seller, who sweetly accepted the 15 dirhams in a way that made it clear I could have another kebab any time. At the station I thought hard about how much I actually needed an omelette before the train arrived. My spirits soared, when it did, for I was going to Marrakech, on a lovely hot spring afternoon. There were swallows again, and, apart from the money, I was all set for my week off. A friend from Wales was coming out to meet me, and we planned to travel together, down into the desert, to greet the main force of the migrating swallows as they came north.

I watched the compass needle spin as we left Casablanca, heading south. There were swallows over the train and I tried to work out which way they were going. Some were indeed heading north-east, as I hoped and calculated they would be, streaking towards the Straits of Gibraltar, but others appeared to be flying aimlessly around the outskirts of town.

It is a beautiful train journey, but still I was a little impatient with it, because I longed to get there, to Marrakech, the town which named the country, and which keeps a part of the heart of everyone who visits her.

Just as we were coming in, and I could finally see the great mountains in their snow, I called my friend to tell him that I was nearly there, and that his reception was assured. He is a teacher like most British teachers, who works too hard for too little. This was to be his Easter holiday and he was looking forward to it tremendously, as I was. At last! Someone to really share it all. He answered the phone and gently gave me dreadful news.

My friend's godfather, one of our teachers, had died. He had been ill with cancer and had fought it, refusing to submit or be daunted by it, right up to the end. A proud South African so naturalised by his work and life to Wales that he actually supported the red shirts when they played the green, he had been in the stadium for the Grand Slam win, and had had, my friend said, a wonderful day. It seemed mad, somehow, but it was absolutely fitting, too. My friend was coming out anyway.

'He would want me to,' he said. 'I was talking about it to him the other day and he was excited for us. He told me a great story about Brazzaville.'

I could hear the shake in my friend's voice and I fought to keep tears out of mine. His godfather had been my tutor at school; he was wise and dry and loathed pomposity and received wisdom: he did a fine line in pithy truth. But there is a hierarchy to grief, as to love; it seemed, absurdly, not to be right to cry if my friend was not. So we controlled it, and said we would see each other soon.

And so arrival in the ochre Marrakechi twilight was not a delight, but a mourning march. My bag never felt so heavy. To save money I determined to walk to a place where I had stayed before; because I was tired I took a short-cut. An hour later I was still making an arse of myself pretending not to be lost. A teenager rescued me and was disappointed with the tip. No one, it seemed, did favours for anyone any more. The line from the rai song kept running through my head:

'*C'est payant, Monsieur, c'est payant . . .*'

Even after my guide left I was still lost, because he peeled away while still some distance from the square, saying the police would bust him if they saw us together.

Finding the place, I obtained a room I could not quite afford. I was resolved not to have my friend arrive to find that we were strapped for cash, having already placed a specific and expensive duty-free order with him, and feeling that it would be wrong for him to spend his holiday worrying about my pennies as well as his own. I suppose what was coming took on something of the complexion of a wake. Sometimes the death welled up inside me, but most of the time I buried it in the life of the streets, the skies, and the bars.

Marrakech was terribly changed. An ugly red tile had entirely colonised the great square, the Jamaa el Fna, like an algae, and spread into the alleys and passageways of the medina. There were neon signs on buildings. There were more of us, with our wallets, and somehow, fewer of them, with their wares, than before. As we thronged the balconies, to watch the famous twilight thickening with the call of the muezzin around the lighting of the lamps, it seemed that we gazed on more of ourselves than perhaps we would wish to see. At rooftop level the change was arresting.

The rooftops of Marrakech are a world apart. As well as offering air

and space and beauty, they are a social and political refuge; traditionally, the roof was woman's place. Risking gossip, admiration and
no doubt the scourge of the self-righteous every time she left the
house, a woman was at least free to feel in possession of herself on her
husband's roof. And, as is only fair, they are the best place to be.
Because they were woman's place, it is the height of bad manners to
stare onto someone else's rooftop. However, because the view is a rosy
and sandy honeycomb of different heights and colours, shapes and
angles, pots and screens, antennae and tables, washing and plants,
birds and minarets, and little half-heard, half-glanced-at scenes of
domestic life, it is impossible not to peep. So although you may see,
you certainly do not look. You may well be glimpsed but essentially
you are invisible. You are in miraculous privacy in an effectively
public place.

It was very difficult not to see people, everywhere. At first I was
alarmed by it. Just two levels above my own and one roof across I
could see the backs, shoulders and occasionally heads of a large mixed
group from Europe, Israel and North America, by their accents. On
another, about 40 yards away, were some fellow Brits. We are
famously easy to spot, at least to other Brits, and for a long time we
thought we had a monopoly on the English Season, as they used to call
the spring in Marrakech. Now, they said, it was an all-year-round
international season. Here and there and here and there, there were
more visitors on rooftops.

'We have high season and very high season,' someone explained,
with a weary grin.

'It's hopeless,' said someone else. 'We've moved to Fez.'

Like many of the riads, the one I stood on was owned by a
Frenchman. With the introduction of the euro a great many of
France's undeclared francs were said to have poured into Marrakech.
Like me, many of the United Kingdom's journalists and travel writers
– no doubt America's and Europe's too – hammered out pieces about
coming here, staying here, buying here, living here, or at the very
least, transporting something of the city's style, aesthetic, cuisine, art
or artefact back home. I have only seen two flight-paths to compare

with the congestion on the way into Marrakech Menara: Barcelona and Heathrow.

But no one who loves birds will ever be unhappy for long in Marrakech. Its rooftops are a wonderful world in which to bird-watch. Not only are you engaged in one of the most sunny, lofty versions of the great human pastime – part worship, part contemplation – that is ornithology, it also is perfectly possible to do it flat on your back on a sun-lounger with a glass of mint tea to hand.

Between the walls of the Bahia palace and the ramparts above the Bab Rehmat cemetery two falcons were in residence. They had a couple of royal palms to perch in, the Andalusian gardens of the Bahia at their feet and miles of tombs and walls to hunt over. They made a lot of noise, at certain times of the day – Kek-kek-kekekekeK! – at other times they were silent; soaring in soft ellipses over the roofs. Below, among and sometimes beside them were the swifts. 'Our' swifts (European Swifts), Pallid Swifts and most of all, Little Swifts, in wheeling crowds of hundreds. For three days, all I really did was worship swifts. But over to the north-east of me, above the Qadi Ayad mosque and the Bab Ailen, were storks, White Storks, and it is easy to be bewitched by them.

A Berber story says they are simply men and women who have taken the form of birds in order to see the world. There is a variation about a woman called Ayasha who left her children to go to Europe, to work at whatever she could for whatever she could get, to make money for her family back in Marrakech. (As is sometimes the way, she was able to obtain the work permits and provide the bank statements her husband simply could not raise.) He missed her so much, and stared at the storks on the royal palace so long, that he became one, too. He took to the air, crossed the great plain, the wide sea and the vast swamp, and followed a mighty French river all the way to the house where she was living. He crashed into her garden, exhausted. And there she was. And there was a Frenchman, and there were children she had had with him. Ayasha took care of the stork, and the more she looked after him, the more she was reminded of Marrakech and the more she missed her husband, and the more the

Frenchman loathed the bird. But he was a good-hearted man, or he cared for Ayasha more than anything, because for love of her he allowed the stork to do exactly as it wished in his house. One day, the stork flew back to Marrakech and resumed the shape of the husband. Returning to his house, he found a great pile of letters that he had watched Ayasha writing, telling him how much she missed him. Not long after that, Ayasha returned, with many presents from the north, and everyone was delighted to see her, her husband most of all. He forgave her adultery: indeed, he never had cause to mention it.

I know this story because a famous Spanish story-teller told it to me; his name is Juan Goytisolo and he lives in Marrakech. (There is a better version of it in his book *The Garden of Secrets*.) He said you can now also hear it in the square, and he is very pleased about this, because it means that an ancient Berber tale, told by one to another, has become a written story in his hands, and has been published for the readers of the world. And since then, no doubt because his telling improved it, somehow it has become an oral story again, and is now told in the square, where stories have been told since the beginning of who knows when.

Have you ever heard a stork sing? Nor have I. Have you seen one dance? Lucky you! But have you ever seen her fly? The wings are set far back; they are broad and strong. A beak halfway between a stiletto and a cutlass leads the way. And her long body and her legs are all held in perfect balance by that white neck, the absurdity that makes sense of it all. When a stork flies it is not that she is using the air to lift her, as so many birds do; it is more that she is using her weight and balance to stop herself being drawn up, and on up, and away.

I lazed about on the roof, watching the falcons, writing up Algeria in my diary, and took siestas in the dark cool of my room in the afternoon. The engagement with the authorities who ran the riad had been swift, bullish and conclusive. For this extraordinary luxury, including wonderful breakfasts and occasional access to the internet, I was paying a fair price. The half-blind, half-paid woman in the kitchen/laundry below me became my friend, and taught me the Berber for swallow: *tififeliste*!

Unlike many birds, swallows call on the wing, on their perches, when alarmed, when mobbing, when courting, when mating, and to one another in passing; perhaps it is another reason we mind about them: they are as talkative as us, and their conversation is as varied. They have a different call for all these occasions. Sometimes, it seems, they just chatter for the pleasure of it. The Berber name, *tififeliste*, is exactly as they sound, when in that sort of chatty mood. Tififeliste, ti-fi-fi-lisss-ti!

Painstakingly, my friend wrote out a vocabulary list, with four keys to pronunciation, and tested me on it, her good eye shining hawkishly, whenever our paths crossed. It was much more fun being a language student than playing my usual role: guiltily tipping guest. I looked for swallows diligently but they were nowhere to be seen. Instead, the world above us belonged almost exclusively to swifts.

Devil birds; flying cross-bows; devourers of wind-borne spiders which sleep on the wing and land only to breed, they are born to the air, to flight. Why do swift-lovers love swifts so? They will give you a dozen reasons. Why do we who love them all divide into swift people and swallow lovers? John Ayto's *Dictionary of Word Origins* gives a clue:

> Swift [OE] The etymological meaning of swift appears to be 'moving along a course'; speed is a secondary development. It goes back ultimately to the pre-historic Germanic base *'swei – swing, bend', which also produced English sweep, swivel and the long-defunct swive, 'copulate with'. Its use as a name for the fast-flying swallow-like bird dates from the 17th century.

The seventeenth century: the age of civil war, revolution, Shakespeare, Milton, the King James Bible and the Copernican shift, when telescopes and astronomy put the sun at the centre of the then known universe, and moved the English and their God slightly to one side. The century of the English language itself: then it was that those of our ancestors who decided that all those swallows were not swallows at all, pointed up at the bands of screaming, keening faster-fliers, and said 'These are swifts!'

In English, at least, swallows are the establishment: swifts are the revolution.

I saw 'our' swifts coming north, and they were magnificent. Compared to the more sedentary species the European swift is a mighty thing. Transcontinental swifts, they should be called, they are like ocean-going airliners, suited in a dark soft brown. Not for them the long, long battle to the furthest south. Swifts fly as though they could go to the ends of the earth, but they have decided where that is: Congo. Its particular storms, its weather systems, its permanently alternating high rain season and low rain season, its vast profusion of creatures, flying, floating and crawling, its morning mists, its afternoon silences, its wild crying darkness, this is the wintering-ground of the swift.

Above the rooftops of Marrakech, however, it was not these birds that beguiled me, but their smaller, stubbier, white-rumped cousins, Little Swifts. The trick is to focus on one bird. To do this you need to be lying down, so that your binoculars, pointed skywards, will not shake. Then pick your Little Swift. To be able to follow one for as much as a minute feels like a real achievement. What at first looks like a batty swarm of dozens and dozens resolves itself into a series of aerial chases. One Little Swift, if not being pursued itself, will almost always be pursuing another.

Locked on to the tail of the swift in front, our hero twists, ducks, dives, slides, skids, arcs and arches, clinging to the track of the hunted. At some point one will either tire or change its mind. If the fleeing swift puts enough space between itself and the chaser, often another bird will drop into the space between them and pick up the chase where the first left off; sometimes the chaser peels away and latches onto another, and pursues it.

A lot of swifty squeaking and screaming accompanies all this activity, and it may have all been in deadly earnest, a mating competition etc., but it looked like tremendous fun. Best of all was seeing them practise their acrobatics. The first few times I saw it I could not fathom it. The binoculars would be left staring helplessly at vaguely unfocused dots: where an instant ago there had been a Little Swift would now be a blank blue space.

In slow motion, then: our Little Swift is hurtling along in relatively level flight. Suddenly it half-closes one wing, folding it in, while half-opening the other; the shoulder moving forward. At the same time, it throws itself sideways and down. It looks remarkably as though the bird has flipped up a hood, pulling the half-open wing over its head. In a split second it drops 50 feet straight down: gone! What a trick. If you can do that your pursuer will almost certainly over-shoot you, and it would not be at all surprising if you did not turn a few heads in the rest of the crowd. Some of them could do it perfectly; others kept practising.

The two days passed quietly while I wished they would hurry up. I ought to have gone out to the Menara Gardens in the evening, or found the sewage works, or at least taken a good long walk around the edge of town, but instead I lay on the roof, and took long, dark sleeps, and soup in the square, and excused myself on the grounds that staying in is the only way of saving money. I thought about what a wonderful way-point Marrakech must be to all migrating birds. All you have to do is survive the desert and the mountains, and in the instant you break over the edge of the precipices, through that lethal wall of winter, there it would be. A bright bowl of smell and noise and colour. All those people and all their waste; what a fug of flies they would put up. How busy, limited and impoverished a species we must seem to birds.

On the evening of the second day a group of swallows burst over the rooftops. They came in fast and low from the south, heading north-east. They skimmed the roofs, barely jinking, at top speed: I had never seen them flying so quickly and directly, with such urgency. I wondered what could be wrong: what drove them? I scanned around with the binoculars and saw, in the quarter of the sky from which they had come, a bird larger than all the rest; the size of a falcon but with something of the speed and profile of a giant swift. It slung down out of the sky in a long, fast glide, a shallow stoop, in fact, and suddenly I realised what it was: the cheetah of the skies – a hobby. Down it came, swift, swift and unmoving, like a missile on a programmed flight path, heading for the great minaret of the Kotubia. And then there was a flicker in its wings, open then shut, very fast, and with a twist it rocketed down at a steep angle, striking, and there came from below the rooftops, just out of sight, an explosion of panicked birds, and it vanished.

It was exhilarating and deeply sinister, like watching a sweep of the reaper's scythe.

'Wow!' I cried. 'Hobby!' and scanned the sky for him again. He was gone. 'A drink is called for . . .' I resolved.

There are two places to buy inexpensive beer in the Medina, which is the old town in the heart of Marrakech. The first is a hole in the wall in the Jewish quarter: down an unlit alley, in a midnight-dark patch where the buildings meet above you there is a hatch, where a hand will pass you the cheapest can of Spéciale Flag in town. The second is the Hotel Tazi.

The lobby, with its sickly yellow light and single other-worldly Christmas decoration high in one corner is a sort of human aquarium. That evening twenty nationalities perched on its tatty furniture and waved desperately at one waiter in an unlikely white tuxedo whose best defence against overwork was to appear to be somewhat confused. Excited Spanish teenagers, German trekkers, loud Italian couples, local bad boys on acid and a host of hawking or touting or drunk Moroccans played musical chairs without music. A man at the reception desk bet me €50 he could guess where I was from, chose Canada, and promised to pay next time. In the dining room, which reeked of cat urine, there had been a buffet supper: all that remained now was a huge, crisped fish head lolling over the side of its dish, its burnt eyes blind to the dirty tables where its flesh was still scattered, its teeth set in a grin. Next door, in the bar, a rank of silently surly Moroccan men watched a football game. The bar itself was deserted but for the harried coming and going of the white-jacketed waiter and a single Englishman from Middlesex who worked with computers, and had brought himself on holiday. It was a pleasure to hear my own language again.

'Oh yeah,' he said quietly. 'It's been an amazing life. Taken me some amazing places. I was in Russia in the 1990s. I remember being taken to a shed in the Urals, just a big empty room and in the middle of it this thing in a big box of wire and wood which was a complete mock-up, which was faking thousands of computers it was talking to somewhere into thinking it was an IBM mainframe! A whole pirated mainframe! Amazing really . . .'

I bought a drink and took it through to the lobby, hoping for a spot from which to observe the proceedings. There was only one seat free, that I could see, just to my right; I hesitated fractionally before taking it because I was suddenly aware that in the ring of occupied places around it, dead opposite it, in fact, was a tall and very beautiful woman. As I sat down someone said, 'Well done!'

She and her friend were in the middle of a comical cross-fire with an aggressively drunk young Marrakechi wearing green. The drunk Marrakechi wanted to talk to the beautiful girl. The beautiful girl did not much want to talk to him, but even more than this, the beautiful girl's no-nonsense friend wanted the Marrakechi to get lost, and told him so, at which he accused her of racism. Switching rapidly from English to French and back again I waded in, determined to take the bile out of the row. It was unexpectedly, unreasonably difficult.

'What for you come to fucking Morocco if you don't like fucking Moroccan peoples?' the Marrakechi demanded.

'Listen,' I said in French, 'it's not that, of course they like Moroccan people, you just haven't understood – leave it, it's not that they don't like you, and they haven't insulted . . .'

'Tell you what, mate,' cut in the no-nonsense friend, who spoke with a dust-dry London accent which reminded me powerfully of Danny, Withnail's drug dealer, fixing the Marrakechi with a stare, 'why don't you just fuck off, yeah?'

'What is it you do?' I had asked the beautiful girl, but in the ensuing explosion from the Marrakechi her reply was lost.

'Sorry – what?'

'I teach communication and peace studies,' she said, and burst into a peal of laughter so mischievous it was almost dirty.

'You are joking.'

'No!'

She asked what I was doing and I explained and asked what on earth she was doing in the Hotel Tazi and she said she was on holiday and had been here before with her estranged husband and her six-, nearly seven-year-old son; she said that it always delighted her. Her soft voice had an accent I could not place.

'Where are you from?'

'Rochdale.'

'Where's that?'

'Lancashire.'

I almost said 'Where's that?' but something happened in my head and so instead with my heart leaping I said, 'My God, I'm completely in love with you and it's really obvious, isn't it?'

She laughed and said, 'Yes!'

(Afterwards she said she had thought I had said, 'Would you like another beer from the bar?')

Her name was Rebecca, and her friend was Rosie. The Marrakechi having pushed Rosie to a certain point, Rosie asked the hotel management to throw him out. This the management agreed to do, but then the Marrakechi wailed in supplication, and begged Rosie to ask the hotel management to let him off, which she did, and they did, and I bought him a beer, and we all finished friends, and two of our lives, at least, were changed.

Enchantment may seem to come like that, like a lightning strike, but of course it does not, normally, no more than lightning comes from a clear blue sky. For years I said I was a romantic, by which I meant that I believed in the powers of life, beauty and art to bring miracles out of everyday existence. By being available for joy, by living in hope and expectation of wonders, I trusted that wonders would come. And so they did. However, I was also – and no doubt am – a man as greedy, venal, lustful, wayward as a man can be, and in the name of this 'romantic' calling I had happily pursued whatever or whoever took my fancy, for years. When it went wrong, when a relationship broke down, I would tell myself that it was because I had not found my soul-mate, and that I must just be patient, and hope. While following swallows it was impossible not to compare their endeavours with mine. What drove their journey, in the end, if not the desire to find a mate, and raise young? What drove mine? It has been said that the business of artistic production is a biological activity akin to the peacock spreading its tail. Was that my motive? Was the pursuit of the birds a pretext for a continuation of my search for the 'romantic' ideal over a vastly expanded territory?

Perhaps. Or perhaps I had outgrown a younger self, or perhaps the journey had wrought some fundamental change in me, or perhaps I had inadvertently been worshipping at an ancient shrine which retained some elemental power: whatever it was, it hit me now like a house falling down. There she was. The one. I have never been so certain of anything. Naturally, part of me thought I must be mad.

Two days later I drove west with my friend from Wales. Rebecca had gone to Casablanca: she would join us tomorrow. I was at the wheel of a rented wreck we had christened the *Petit Voleur*, a pre-scratched Ford, elegant as a bread-bin, battered as a skip, hired from no reputable firm, which pulled viciously to the right, with a slaughtered gearbox, which carried Casablanca plates and screamed 'Stop me!' to every policeman.

Beside me was Norddine, an acquaintance of Rebecca's, a young Marrakechi, a walker, a musician and mountain guide, with a sitar between his knees; in the back were Rosie and my friend from Wales. In the boot, a pile of treasures belonging to the Two Princesses, *les Deux Princesses*, Rochdale girls of Pakistani descent, who were travelling with Rebecca. With their Punjabi, their great beauty and iron-hard bargaining, they brought first hope then despair to every trader whose wares attracted them. They bargained like pirates. We were carrying their haul from the souk, and now following the westering sun as it sank, casting straw-gold light over the ochre lands which led, in the end, to the sea.

We were pulled over by two unsmiling officers of the Gendarmerie Royale, grey-uniformed police. They asked for a demonstration of the brake lights, almost as though they knew they were not working. I was outraged, having hired the car only hours before. They summoned me to their vehicle.

'The lights do not work. You must pay . . .'

'Yes! I will!'

'What?'

'I am more than happy to pay. I am absolutely furious – it's incredibly dangerous not to have working brake lights.'

The officers were entirely wrong-footed by this. They began to smile.

'How much is it?'

'Where are you from?'

'Wales . . .'

They nodded, and continued to smile.

'How much is it?'

'How much do you have?'

'Well that is the problem. I spent it all hiring the car and paying for the petrol – I need to go to a cash point in Essaouira . . . I can't believe the bastard who rented me this car!'

By now they were laughing with something like delight and shaking their heads that anyone could be so gulled.

'So how much do you have?'

'Fifteen dirhams!'

They took it as the most enormous joke, accepted the fifteen dirhams, still shaking their heads, and let us go. It was as though something was on our side.

Perhaps something was on our side, that weekend. We made the crest of the cliff above Essaouira just as the sun sank into the freezing

Atlantic and the sea turned spray-white and blue. Essaouira's winds bit us and we hurried into town, and huddled over coffee, and Norddine found us an apartment. Essaouira is defended by its winds. An ancient trading post, white-walled and ramparted by the Portuguese, it has a harbour, guarded by the Isles of Mogador, and a beach which stretches as far as you can see down to a castle slowly being eaten by waves. Were it not for its winds Essaouira would have been bought up long ago and Norddine's find, a flat in the centre of town, with a balcony over the street, within earshot of the sea and a roof with a view of everything would have been owned by some distant millionaire and entirely unaffordable to us.

The next day we ran on the beach, throwing ourselves into the wind, and sitting on it, and my friend and I re-enacted the jinking runs of the Welsh winger, Shane Williams, who had scored for us the winning tries of the championship. We watched a man kite-surfing in the ferocious wind and my friend noticed that of the six camels available for hire on the beach, two at least were so camp that they had to be gay. And that night we feasted, when *les Deux Princesses* and Rebecca arrived, and we drank beer and whisky and argued about abstract things for pure pleasure, and told stories, and bargained for my hat.

'There are only five beautiful girls in Rochdale,' they said, laughing, 'and you've got three of us here!'

Norddine, my friend and I felt appropriately blessed. My friend told the story of *2001, Space Odyssey* and I felt I was watching it. Between stories we haggled for fun; we worked out the bride price of one of *les Deux Princesses*: a girl, by common consent, extraordinarily pretty.

'A BMW X5 with tinted windows and a full tank of gas,' she conceded.

Then Norddine told the story of an adventure in the High Atlas.

'I went for a walk for three years: I walked the mountains and beyond the mountains. I saw many things, I met many people, I learned many things in that time. One day I was walking in the mountains with my friend and my dogs. I have a bow – do you know

archery? – a proper bow. You should see it. Anyway, we were far away
from everywhere, somewhere very remote and very high up, and in
these mountains there are apes. Normally my dogs are very good, they
do what I tell them, but that day they chased some apes – I called them
but they didn't listen and they chased. So, we carried on going, and
the dogs came back eventually. Then my friend said – look, up there,
apes. And there were some apes high above us. The dogs barked at
them and then there were more, on the left side, and more, on the
right. We stopped, and turned around, and there were apes behind us.
We were in a bowl in the hills, like a bull ring, and there were apes all
around us now, hundreds of them, and they closed in. And my dogs,
you should have seen them, they started to tremble, shaking all over,
and pressing against my legs, and still the apes came closer, and they
were barking and growling and banging on the ground. My friend
said, "Hey, we've got to get out of here – they want to kill these dogs!"
And the dogs knew it. They were shaking so much and whimpering,
you know what, they pissed themselves in fear. My friend said, "Fuck
this! Come on – if we don't give them the dogs they're going to kill us
too!" But you know what, I had seen the leader of the apes. He was a
little way back, and a little high up, and he was big, a big pale blonde
ape. And I thought no, you're not going to kill my dogs, so I took an
arrow and I shot near the closest ape, but they still kept coming so I
took some more arrows and I shot a few apes, each time getting closer
to the leader. And when he saw this, something changed, and the apes
stopped advancing, and a way opened behind us, and they let us all
go . . .'

I did not disbelieve this but there was something strange about it,
supplementary to it, as if Norddine's story was a parable.

My head span with the mysteries and tricks of Morocco. The way
an ashtray is just two pieces of clay which fit together – click! –
smothering smoke. The way a key may be turned in a lock, round and
round, snick, snick, snick – uselessly, until it is turned by someone
who knows how to use it. Then the door opens. The way they have
with a credit card and a pencil, shading the waxy paper foil the way a
child traces a penny. The way signatures are smeared, blurred, made

meaningless. The endless maze of SIM cards, changed phones, changed numbers – as if half of society has evolved in deliberate, willed obscurity, hiding its doings from the police state. The way a card reader can be fixed so that it takes twice, three times, four times the amount you keyed in, in seconds, humming like a saw.

'My brother installs satellite dishes,' someone said. 'He has a satellite decoder. You can scroll through satellite feeds like tuning a television!'

'Do you see things?'

'You see amazing things . . .'

In the market place Bush and Bin Laden chase each other around a toy train track, grinning, never catching up.

'He's dead you know,' says someone, seriously. 'He was ill – he died two years ago.'

Norddine teaches me a password: Azamir. Azamir! you say, and doors open, miraculously. It is the Berber for Berber. You soon realise that Morocco is a nation containing a nation, the Berber people and culture follow the Atlas right across North Africa, from the Atlantic to Algiers. If you could speak their language you could move across borders, through unknown worlds, like a ghost.

'Ha! The Welsh are the Berbers of Britain!' I declare. 'We are hill farmers too! Azamir!'

The house was like a lantern of winds that night; the gulls laughed and cackled and cried like spirits in the luminous moonlight – houaa! houaa! – and just before daybreak the muezzin called, loud and as if forlorn – Allah-u-akbar – Allahuuuakbar!

We went down to the port and the dockyard where fishing boats are still built. We were given a tour of the yard; they showed us the different woods that go into each boat: teak, and, in the crucial joint at the point of the keel, the piece that holds the whole boat together, iroko, from the plundered forests. We lolled on the dock and told sea stories. The wind had stopped, as if by some sorcery, the blue swells of the sea seemed to check themselves and the air was as still as held breath. After lunch I bargained for a silver ring for Rebecca and a strange cross for my friend, who is a mariner: a Berber compass, for finding your way in the desert.

Another of my teeth blew up. It went without any pain: there was just a crack in my head, from the upper left side of my jaw, and bits of molar fell out, leaving a good-sized hole. I rushed for my dental kit, but there seemed no call for it. I left part of myself in Essaouira.

In the evening I took scraps up to the roof for our two nesting gulls; it seems many Essaouira residents do this, as if making offerings to djinns. We went to a bar overlooking the ramparts where ancient Portuguese cannons still point out to sea, and drank, and danced. A strange, shifty boy talked to me about Essaouira. Did I know it was all run by a Jew? No, I said, I did not.

'I used to be banned from here, but they let me in now, because I have changed,' he said. Then he asked me for money. He did not seem to need it but I felt – in the flood of love and hope that embraced me, with these new friends, and my greatest friend, and this laughing, beautiful woman whom I felt I had always been looking for – as though I had no need of money either, and I gave him a note. My friend did not understand it.

'But why would you do something like that? I don't understand.'

He was offended by the waste.

'Because if he wants something so badly it doesn't hurt me to give it.'

It was not the first thing I had given away, but it was the beginning of a flood of giving, a kind of divesting, which came on me and would not stop. As we were leaving the bar's manager caught me hiding money under the ashtray.

'But why?' he asked.

'Because I used to collect ashtrays, once, in a club.'

'Would you like a job here?'

'Yes!'

'What can you do?'

'I'm a barman.'

'OK,' he said, after a short pause. 'Come back one day.'

Norddine and I prepared a feast as if we were catering for a wedding. Later we went up to the roof and sat with the guardian of the house. I had a piece of green glass which I had picked up on the beach.

'What is this?' asked the guardian, turning it over in his palm.

'I found it on the beach.'

'Yes, but what is it?'

He ran his fingers across its edges; some rounded and sea-smoothed, some still sharper than razors. Its shape reminded me of a country on a map.

'It's . . . Algeria!' I cried.

He laughed. 'This,' he said, turning it over again and then dismissively tossing it on the table, 'this? This is life . . .'

In the morning my friend and I bade farewell to Norddine – I gave him my precious head-torch – and to Rosie, *les Deux Princesses* and Rebecca. The girls were going back to Rochdale and London – half-term was ending for them – and we were going to do a great circle, south, first, to look for swallows coming in from the desert, then north again, to Tangier, the Straits and Spain. We pointed the *Petit Voleur* at the Anti Atlas and drove.

We did not travel well together. I was over-excited. My friend became very quiet. Perhaps I had been on the road too long to share it reasonably. We were together but it was as though we saw different worlds. I have nothing but strange memories and questions from those days. I am only half-sure of the memories: perhaps my friend could confirm them but he will not; it will be years before we reminisce about Morocco, if we ever do.

Did we really go down to that strange beach, in pewter sea-light, and help the men push their fishing boat through sand like a crust of clay, and swim, and imagine we might live there? What was the name of the boy with dreadlocks whom we ate with in Agadir at the fish market, where we drank surreptitiously with the off-duty policeman? Where were those orange groves where we stopped to do nothing, and all the birds singing? Why did we buy carpets in Taroudannt? Where

I gave away my hat, my beautiful, treasured, road-battered hat, to the old bald man on the bicycle. And then we met Brahim, a guide, a trader, a clever and warm but somewhat mysterious man, and read that cutting he gave us, too, written by an American – was it really from the *New York Times*? Did the writer not say that through Brahim he had received a warning about a terrorist attack on America? Why did Brahim show us that? What was he trying to say? Did he think we were messengers?

I split my head open, playing hopscotch on black and white tiles, and my friend bandaged it. Was I concussed, in the following days? Would that explain the strangeness?

Brahim took us on a tour, a long, wonderful drive through the Middle Atlas, past the king's gold mines, all the way down to the desert. We stopped in the darkness, in the sand, to meet Ahmed, a man of infinite gentleness, and two camels. We rode, perched on our camels, south-east, through the dark dunes. And what were those shapes that shadowed us? Jackals? Djinns? Mujahadeen?

We counted meteorites that night, as we talked and argued and cried, under a sky that seemed to burn and pulse with stars like flecks of fire. I went out, later, while my friend slept, with my camel, which refused to carry me to the top of the dune. I saw the moonrise over Algeria, just after dawn, huge and perfectly round, like a ghost, like the sun's dead twin.

Out there, beyond the encampment, there was no way to judge distance, height or depth. With nothing to fasten on, the eye assigns uncertain values to the landscape – that is far, this is near, everything in between is a speculation. There are tracks on some of the dunes – fennec foxes, gerbils – their marks the only tiny traces of life between the sand and the high, bare sky. Under our cloudscapes, we of the north live in a narrow strip which allows us to scale the earth according to our own proportions. A man is so high, a tree is higher, then a building, then a hill. But in the desert there is no such relativity; here you are confronted by space the eye cannot measure and the mind cannot calculate. To climb to the top of a mountain in Britain is to be exalted, as the land spreads below you for your contemplation.

But to gain a vantage point in the desert is to be confronted with the daunting exaltation of untamed space, to be further diminished. Your sense of self shimmers at the sight of the dune sea. The tenets with which you armour yourself against existence seem vain illusions against the desert's beauty and indifference: you are a tiny, tiny, temporary thing, fragile and vulnerable as a bird. To stand in the dunes, a speck on the sand, is perhaps something like finding yourself in the water, mid-ocean, beyond the sight of ships and shores. To find my way back, not trusting my tracks in the sand, I left one shoe, then another, as markers on top of high dunes.

The desert, more than anywhere else, raises the question of how swallows find their way. Ornithologists believe that there is no single answer. Experiments have shown that they are sensitive to magnetism, perhaps able to see, or at least sense, magnetic field lines, which allows them to distinguish between two directions: towards the pole, and towards the equator. However, the same experiments have also shown that they do not use this compass when the sky is not overcast: in clear conditions, it seems, they orientate themselves by the position of the sun. It is also thought that they use landmarks, and perhaps olfactory cues: navigating by the smell of the terrains they pass over, and certainly, it is thought, the distinct scent of their breeding grounds.

Ahmed taught us to read the winds. The Chirugui blows from the east, just before dawn, the north wind blows at night in that season, and the south wind, the Sahel, rules the day.

In the morning we rode out of the dunes into a sandstorm, a yellow-white fury which thrashed the palms and cast an unholy pallor over the daylight, it was as though the world had aged and paled into an antique photograph.

We took turns driving Brahim's jeep, a thing so sophisticated it seemed to me almost a weapon. And the mountains were so cold and so high, with icy lakes where I saw swallows, and the ridges were coiled like burning black and copper-red monsters: terrifying, awesome, bared and shaled and sharpened like dragons' backs.

In the morning in Fez I was on the roof watching swifts and falcons and Pallid Swifts, and then we set out again, after coffee – the only vehicle on the road. And what were all the police waiting for? I was still giving away – I gave my binoculars to Brahim's niece; my eye was now so trained I could spot swallows without them, and when the girl said she was interested in birds it seemed that she would gain so much more from the beautiful device than I ever would. I felt a great tiredness at all my possessions and equipment, something like a longing to be free.

Everything seemed better, as we neared Tangier, through the rain, and we rolled down the windows, and the world smelled salty, green and sweet, and we exulted because it smelled exactly like the warm spring morning on the coast in Wales.

We hugged Brahim and ran pell-mell through the port, to our ferry, and scrambled up onto the deck, and my friend was delighted to see swallows:

'Woah! Look at them all! That is quite wonderful, actually . . .' he said, because they were all around us. Cutting through the drizzly air like little racing boats, flinging themselves away from Tangier, out of Africa, low and furiously fast across the Straits, they headed for Europe and we went with them. Our boiling wake seemed to tow the low coastline after us and I had a waking daydream that the Straits were closing behind us, and imagined Hercules drawing Gibraltar and Jebel Musa together again, closing the gap that splits one world from another. Not long ago they were finding over a thousand bodies a year washed up on the beaches of Morocco and Spain. No one knows how many drown and disappear, unfound, unclaimed.

CHAPTER 10

Gibraltar to Madrid: The Rock and the Line

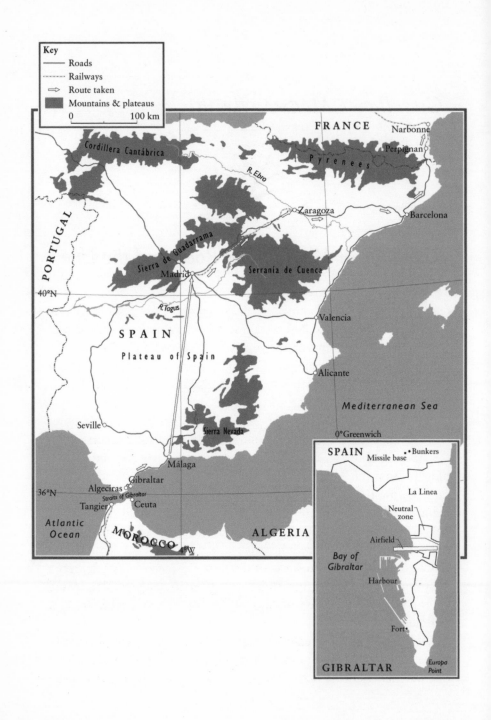

Key

Gibraltar to Madrid: The Rock and the Line

When I planned the journey I made glib assumptions about crossing from Africa to Europe. The swallows, I hoped, would teach me what it is to travel like a migrant, and to contrast real boundaries, jungles and deserts and seas, with notional ones: the lines that humans have drawn on the world. I will arrive in Europe with my eyes peeled, my sight renewed, I thought, I will see our continent as if for the first time, the way immigrants from Africa see it. I had no idea what that would mean.

At their narrowest point the Straits of Gibraltar are only 8 miles wide. From the Moroccan side you can see wind turbines turning on the Spanish shore. It is a place of wild weathers: fronts and squalls blow in from the Atlantic; brief and fierce sunshine is veiled in gusty rain. Below the surface, seven currents interweave, the principals being the Mediterranean outflow, which is highly saline, and sinks below the incoming tides of the Atlantic. The ferry heads east, travelling 30 miles between the two coasts, aiming for the semicircular Bay of Gibraltar and the docks of the port of Algeciras. Algeciras is a busy and crowded port, but all its activity is dwarfed and dominated by the massive upthrust of lion-yellow stone that is the Rock of Gibraltar.

My friend and I parted after customs at Algeciras. He planned to take a train to Barcelona; I took a taxi to Gibraltar. It is famous among birdwatchers as being one of the great places to watch migrants, and I was keen to see it, this little outpost of Britain on the Mediterranean. The driver was a Catalan with no English. Our communication was poor but he insisted on teaching me one word. Not 'Gibraltar' but 'Yibralta', he said, and made me repeat it.

Beyond customs was a red telephone box; beyond that a cash machine poked out Gibraltar pounds like paper tongues, decorated with fighting ships. My book advance had come through: I was solvent again. In the streets people acted differently, walked differently. I could not define it at first but then it came: it was in the eyes, in the gaze. A passer-by in Europe carries a distinct expression, a certain look, as though the stare is not a reaching out of the gaze, but a by-product of a turning in. As though existence is self-consciousness, as though the world itself is not a curiosity, far less a wonder, but a distraction from the essential business of the self; as though the life of the street is humdrum, something to be blotted out.

The high street was like something out of the home counties, with signs trumpeting VAT-free shopping; I walked under surveillance cameras and everywhere passed plaques commemorating victories, deaths and defences. In Gibraltar guns and ships and flags are fetishes; high up, a Moorish castle flying the Union Jack now serves as a prison. Men walked with bulled shoulders, their hair cut to fit under berets. The Rock itself bristled with aerials and dishes: the Rock was listening; everything was watching and listening, yet nothing made noise. In Africa music played all the time, everywhere: here it was silenced, piped through headphones; one of humanity's great communal experiences was here used to isolate, to create private experience. Europe's shared music is the sound of traffic.

In the foyer of the hotel a *Daily Telegraph* and a *Daily Mail* lie like flags of allegiance on the counter at reception and the staff call me 'sir' coldly, because I am thin and brown and a bit dirty.

The hotel is decked and panelled like a ship. I withdraw into the cabin of my room and revert as if for comfort to my African habits, washing my clothes with soap in the sink. Night falls and all the lights of the bay come up; ships move in slowly, buoys flash, docks are bright under arc lights. I hang my clothes in the window to dry; in a time of war I would be accused of signalling to the enemy. Gibraltar feels as though it is on a war-footing, so ubiquitous is the presence of the military. One of Britain's most infamous peacetime executions took place here, on a street in broad daylight. Three unarmed members of the IRA were shot down by the SAS. Eye-witnesses who said the soldiers gave no warning were smeared by the newspapers, and the government tried to prevent the broadcast of a film about the incident, *Death on the Rock*. There is nothing innocent about Gibraltar; there is no peaceful or civilised explanation for why it exists, except the claim of all powerful civilisations, that they prepare for war to preserve peace. Gibraltar is a strategic asset, a weapon.

I did not eat but I drank. I lay down and dreamed but did not seem to sleep. Late, I got up and went out, my bag packed, resolved to leave. I climbed a stairway from a courtyard behind the hotel onto a roof. With binoculars you can watch the migration at night by focusing on the moon. If a bird flies across it you will have a second in which to attempt identification. But I had no binoculars, so I lay down under criss-crossed washing lines and gazed up at the Rock. I saw the spark of a cigarette lighter; soldiers, an exercise, a watchman? I slept a little, then got up and went south on silent roads. A patrol boat seemed to keep pace with me. I felt watched again. I followed signs to Europa Point. Something was growing in me, like an anger or a madness. What was this place? What was it watching for? What were you to make of this strange little British fist, clenched around stone at the mouth of the Mediterranean? Home? This was not home, this was a garrison. I abhorred myself for having scuttled here, for taking refuge here, with my little sack of treasures, and my notebooks, two of them, bursting with records of all that I had seen and felt. I thought back to the Kovango, to the pages I wrote – how does the white man come to Africa? Like a treasure hunter, above all, seeking reward – financial, spiritual, experiential. And like a treasure hunter I was suddenly 'safe' in a fortress, under the Union Jack again, watched over, or just watched (the patrol boat still seemed to keep pace with me), 'safe', 'home', a 'citizen', a 'traveller' 'returned'; mission accomplished. It made me feel sick, this parody of completion in this parody of a country, this fortified Little England.

I walked faster and faster through rain-salted darkness until I came to the end, to a crashing black void of indistinct waves below the cliffs and a single lighthouse waving its white arms at the sea. Without thinking or hesitating I slipped my arms out of the straps of my rucksack, grasped them and began to spin round and round on the cliff top, round and round like a hammer-thrower, until the bag seemed to have a lift and velocity all of its own and I snapped round one last time, jerked my arms up and let go, sending it all flying, everything I had collected, carried, preserved and noted: I flung it all off Europa Point into the darkness, and it rose and then fell, tumbled down spinning, into the body-hungry sea.

Perhaps a doctor would have said I was suffering from stress, exhaustion – I had not slept much at all, through those nights in Morocco – malnourishment (I was 2 stone under my usual weight) or even concussion. But I felt better, as I walked back to the town, and the light came up, and Gibraltar began its new day. Soldiers drove to work as young men do, all roaring engines, and in the barracks on either side of the road others came in from a night exercise and made themselves breakfast. I realised I had left my passport and wallet on the roof where I had slept so went back to the hotel to fetch it. At the hotel they said that roof was nothing to do with them. Then the police showed up: a tall Englishman, a young Welshman and an Irishwoman police officer, and they were very civilised.

My passport and wallet had already been handed in, and the police wanted to know what I had been doing on the roof in the middle of the night. I was intending to watch birds, I said.

'Very well,' said the English policeman. 'There you go. And if I catch you climbing on any more roofs in Gibraltar I'll arrest you – all right?'

'Fair enough.'

I walked out of Gibraltar, back into Spain, and followed swallows through the streets of La Linea.

I still had the skeleton of my phone but the battery and charger had gone into the sea with my rucksack. Travelling for the first time without communications, compass, maps, soap, toothbrush, razor or a change of clothes was nothing compared to travelling without language. La Linea, a small town strung along the coastal strip in the hinterland of Gibraltar does not welcome English or mime. What Spanish I had became confused with Italian, which I speak a little. For the first time in the whole journey I had no tongue.

I must have seemed strange; La Linea is not a tourist town: it was the first time that I had been looked on not as a traveller, but as an unknown quantity. People's expressions were not friendly. I found myself in a supermarket, confused and dazzled by the huge choice of bright goods. I asked for a coffee in a bar and met outright hostility. I do not know why, but they threw me out, a young man angrily shouting something I did not understand. I lost my sense of direction and wandered in circles in the thickening rain. The only thing I recognised were the birds: there were swallows cutting through the low streets, slicing rapidly across the grid of small yellow houses.

On the outskirts of town was a stretch of wasteland and beyond it a rise to a stand of trees. A path led past a junkyard, up through the trees and on, past tethered horses, to a hill. High ground would give me bearings. I went slowly now; the soaked leather of my shoes rubbing sorely against my heels, my legs heavy. I had been walking since four in the morning. The rising ground turned in to the foot of a wide hill, climbing away from the sea to the north, covered with spring flowers, spiky bushes and hardy, succulent plants. I took my shoes off to give my blisters a break. The rocks were sharp and the bushes wet with rain. I came to a concrete pillbox, and then another. From the top of one you could see many more, all supporting each other, dug into the flanks, ridges and crests of the hill. They were all serviceable; I half-expected them to be occupied. What were they doing here, with their gun-slits facing the Straits, and Africa? Anywhere else in Europe I would have assumed that they were relics of the Second World War – but Spain had been neutral then: what army were they supposed to oppose? There was something horribly sinister about the network of pillboxes and slit trenches; it was like stumbling across the preparations for an undeclared war. 'Fortress Europe' was suddenly more than an expression. It had real battlements, real teeth.

Each ridge, surmounted, revealed another beyond. The pillboxes climbed with the contours: at the top of the hill was a gate and another cluster of fortifications. There was an army truck there, but no sign of anyone. Curtains of rain swung in from the north-west. Below, beyond Gibraltar, the Straits were a pearly swirl of changing weathers in which the coast of north Africa came and went. I reckoned I had an hour left in my legs before I would have to stop. The way forward was barred; the nearest road was below me to the west. There was a fence down there, but even from here a wide gap was visible, and not far beyond that a road and traffic.

It took half an hour to descend the hill and reach the gap in the fence. It had been a formidable construction, once, but was now rusted and sagging; from the plant growth the gap had been there a long time. It would save a considerable detour. Not far beyond the fence was tarmac.

The first sign of trouble was another army truck parked under some trees. My heart sank a little but I continued, too tired to turn back. The next truck was an unfamiliar design, with a rig at the back and stabilisers. The road turned a corner and I found myself walking between a double line of trucks. There were missiles mounted on their backs. The eerie thing was the silence. There was no sign or sound of anyone, anywhere. The rain had stopped; it was a still, strange day and a strange moment in it, neither afternoon nor evening.

There were a great many more missiles on either side of the road. They did not look very modern, more like antiquated, Cold War things, but I was scared to look too closely. All I could think to do was to keep staring at the ground, to not see too much, and keep going through a large and deserted barracks like a film set, like an aftermath, like a scene of the end of the world: trucks, and missiles, and silence. I suppose I thought a combination of innocence, error and being European would be some protection.

Beyond an open space like a parade ground was what looked like the main gate, with a barrier and guard hut. I walked towards it very quietly. I could not see anyone. I began to walk past it, practically on tip-toes, and was at the barrier when the shout came. There was a woman in uniform in the hut, summoning me, a puzzled expression on her face. I thought about running but did not.

She asked questions I did not understand. I tried to explain that I had made a mistake but nothing I said made sense to her. Then she asked, in hesitant English:

'Are you the man who is running?'

I did not know what to say. Three soldiers approached, carrying rifles; their officer was another woman. More questions, more incomprehension. Then another officer appeared, with a small orange dog. Perhaps it had been trained to look for drugs: the soldiers hefted their rifles and everyone watched as the dog sniffed me. Another soldier came, in camouflage, wearing a sludgy green beret. He seemed different: by his appearance he could have been English but he did not respond to my appeals. Was I the man who was running? Had they mixed me up with some sort of escape and evasion exercise? I hoped not, or it would be interrogation next. They were waiting for something. Gesturing with their rifles they herded me away from the gate towards a doorway.

'Now what? Up against a wall and shot?' I asked, not entirely lightly.

They did not respond. The barriers opened and a four-by-four sped through, blue lights flashing: enter the Guardia Civil. There were two of them, a tall youth and his boss, a short, portly officer with thinning hair. The latter conferred briefly with the soldiers, then advanced, shouting questions. I am a writer, I am watching birds, I came through a gap in the fence by mistake, I tried to explain, but he understood nothing and became angry, screaming questions into my face. His anger seemed to swell and ebb in waves. One moment he was calm, the next his hands shook and he grabbed me. He was sweating. There were beads of it on his scalp and brow. He threw me against the wall, pulled me back, turned me round and threw me against it again. He kicked my legs apart and spread-eagled my arms. There was a fierce, almost sexual excitement in him. Before he moved in he glanced, almost involuntarily, over his shoulder, exactly like a school bully checking that he is unobserved. It seemed to confirm that this was off-the-record, beyond control or formal process. Another blow came, a sort of cuff. I was repulsed but not seriously frightened: had I not been white, had I been female, had he had me in a cell . . . Perhaps nothing more would have happened, but it was horrible to imagine.

Then, for no reason I could discern, it all stopped. The officer spat something at me, spoke with the soldiers, and I was hustled to the barrier and pushed past it. I walked down the road very slowly. At the bottom was a roundabout where I stood with my thumb out for a while, until a man stopped. We did not understand each other but he drove me into La Linea again and put me out in front of a tall building, the Hotel Ibero-Star.

In normal circumstances this hotel might have seemed unremarkable but that evening, in that state, it was repulsive. It was the most modern building I had been into in weeks, equipped with everything today's business traveller presumably expects. It was entirely devoid of personality. There was no sense that the staff could make any decision: their role was to operate the hotel like a machine, to apply the rules, to uphold the corporate image, to carry out the corporate function.

The cleanliness was inhuman: the whole place was biologically, forensically, hysterically clean. This fortress of sanitation seemed designed and determined to resist the grubbiness that is people – I was certainly the grubbiest it had allowed in for a long time. The manic, authoritarian air of hygiene frightened me; the bathroom was terrifying. A basket of toiletries sat on a ledge like a surreal sculpture of a crustacean. The facecloth had been coiled into a crescent, its corners forming claws: a shell of soaps and shower caps made a body and shampoo bottles with spherical caps protruded from the front like crab's eyes on stalks. It was such a baroque, extravagant construction that I shrank from it. This must be policy; there must have been one in every room. The shower was so clean it seemed never to have been used. The entire room seemed to be wrapped in a layer of invisible cling-film. On television men in suits discussed something they called a worldwide financial crisis. The crisis did not seem to be hurting them: they laughed at it, shook their heads and shrugged in smiling disbelief. The glitz of television made it all entertainment, the rolling news channels rolling everything smoothly into the ad breaks. A card on the desk promised rewards for loyalty, trumpeting the virtues of the Ibero-Star chain. Where we spend, what we buy, to which corporations we are loyal – these things will displace passports, I thought, staring at it, corporate loyalty will define us more powerfully than the mere accident of where we are born, as long as we are born into the fortress.

Breakfast was the same in its extreme fecundity, a jewelled wealth of bright wrappers, gleaming pots, everything advertising itself, a banquet of edible marketing. Afterwards I stood on the pavement, at a loss. A man pulled up in a people-mover.

'Are you waiting for me?' he asked. He was English. He seemed friendly and assured.

'I don't know – who are you here for?'

'I'm supposed to pick someone up and take him to Malaga.'

'Yes,' I said, for no reason.

We listened to Michael Bolton on his stereo.

'As I've got older,' he said, 'I've come to appreciate this sort of music more – you know, just nice music, nice words . . .'

He dropped me off at the airport. I went in and bought a ticket to Madrid. As I passed through the scanners, at the point where you empty all your pockets into a tray I was struck by the impulse: with a British passport and a wallet full of plastic you can still go pretty well anywhere, replace pretty well anything, pay at any door until it opens. I walked away from the scanner without picking up my tray, abandoning everything to the X-ray machine. I landed at Madrid with a few euros and the clothes I was wearing. Of everything I had set out with, from my birth certificate upwards, nothing now remained.

It was a cold night in early April; the streets gleamed after rain. I hung around the bus station for a long time, keeping warm, then started to walk. Madrid seemed vast and empty, long pale boulevards rolled away, exhaustingly, in every direction. It all seemed so sudden, so established and indifferent that I felt I had fallen into a new world, deserted and austere. I asked a policeman if he knew anywhere cheap to stay. He asked me for my identification. I told him I had none. He said I must report the loss of my passport to a police station. What police station? He showed me on a map. I looked for an hour and I must have been close, at times, but I could not find it. A boy who was putting up posters gave me a lift in his van at one point. In the small hours of the morning I found the police station and was told to wait. A girl was waiting too; she was feverish with excitement.

'I'm going to report him this time,' she said, in English. 'I'm going to get him arrested.'

'Who?'

'My neighbour, the bastard.'

'Why?'

'He won't leave me alone! And this time he went too far.'

'What did he do?'

'He banged on my door, shouting that he loves me.'

'You're going to get him arrested because he loves you?'

'Because the bastard bangs on my door!'

The police took my statement and gave me a form.

'Now go,' said the officer.

'I have nowhere to go – may I stay here?'

He looked me over. 'Two hours,' he said, pointing at the bench in the waiting room. 'You can stay for two hours, but don't lie down.'

I closed my eyes. Immediately, it seemed, though two hours had passed, they shook me awake.

I walked through the streets as the sky lightened. The air was fresh and cold. There was a bewildering treasury of waste on the pavements. In one skip were jumpers, a bag, shoes, umbrellas, shirts, jeans, shorts and T-shirts. I re-equipped myself, because such riches seemed too good to pass up, but the weight of the bag was exhausting and I returned it to another skip. Madrid's skips and recycling bins act as an exchange for street people: you trade in your rags for anything better you find. I counted my euros: five left. A café owner let me use his bathroom.

'Do you want something?' he asked, as I was leaving.

'I have no money,' I said. He shrugged, made me a coffee and presented me with a heart-shaped cake on a plate. His face was stern and unforgiving, but his kindness was amazing. He showed me the door as soon as I had finished and waved away my thanks, almost irritably.

People were going into a church for morning Mass. I joined them, took a seat and tried to stay awake. The Latin was soporific and I failed to keep my eyes open, awaking to a hand shaking my shoulder. A man stood over me, dressed in ragged clothes. His head was swollen and his eyes bulging. His breath came in loud, whistling snorts. He seemed to be the self-appointed guardian of the church and he was upset and angered by me. He chased me out with shooing gestures, huffing and whistling furiously. I walked through parks and down long streets; I huddled on benches. My head spun. At one point I went into a hospital but they did not understand what I wanted and were not interested in the cut on my head. I was neither tourist, nor traveller (whoever heard of a traveller without a bag?), nor worker, nor resident: I felt as though I had fallen through a grating into a kind of invisibility. Madrid seemed to divide into well-to-do quarters, where people looked through me, and bohemian areas, where entire squares were filled with colourful crowds of rockers, bikers, beggars and students – the anarchic and the bourgeois separating themselves like oil and water. The astonishing thing was the way this city, the

European capital closest to Africa, did not refer to that continent at all. Instead, it was the tidal pull of another empire, another history, which was everywhere: South America. I gazed at Ecuadorian, Venezuelan and Peruvian faces, music and food. I saw some Moroccans, but their culture was comparatively invisible. I spent another euro on a coffee. Cigarettes were no problem; every ashtray was rich in butts. It began to get dark again and I joined the hurrying crowds, to keep going and keep warm, but also to feel part of the city by appearing to be, to keep its rhythm. Two euros. There was so much food in the cafés and shops but I could not work out where they threw it all away. I was desperate to sleep somewhere. Finally, I went into a hotel lobby, borrowed a phone book and found – miracles! – the name of the parents of a friend. One euro went on that, and one on the metro ride to the station near their home, and that is all it took. One phone number, two euros, and the kindness of near-strangers.

I felt pathetic as I made that call; the equivalent of calling your parents, crying 'Make it stop – help!' As a privileged European it was, perhaps, rather feeble, but it was such a relief: I could have sobbed with gratitude when the telephone was answered. For a migrant, it would have been a God-send. As I discovered in the following days, the telephone number of someone in Spain is the one life-line possessed by many of the 'clandestins' – as they are called in Morocco – who make it across the Strait.

Judy, the lady of the house, came out to fetch me from the metro.

'Where are your bags?' she cried. 'Have you lost them? Have you eaten?'

We had met before, fifteen years ago: I attended the same international school as her daughter, a close friend. Judy and her husband Denis had become used to various friends of their children turning up in various states, but Judy looked alarmed at the state of me. She fed me stew, took my clothes away for washing, lent me a T-shirt and jeans, and showed me to a room with a bed with clean sheets.

It is hard to understand how precious and wonderful are food, clean clothes, a bath and a bed until you have been given them, with such kindness, in such need.

The morning was as bright as sun on snow. The light was luminous, the sky a freezing blue; in the suburb where my friends live there were pine trees and occasional sudden views for miles.

'It feels like the first day of spring,' we said.

Denis and Judy seemed to take me in hand in a very gentle way. They looked closely at me, asked about the journey, interrogated me as to my plans and resources, gave me a telephone, making it clear that it was time I called my parents, and declared that I had not been eating enough.

There were swallows on the wires over the road and a wide open scent of cold freshness, like the air of the mountains, and in the lancing sunlight every leaf seemed distinct and every scent was strong: Madrid woke with coffee and tobacco and baking.

'We had a swallow,' Judy said. 'A beautiful little thing, he fell out of his nest and we fed him; he used to drink from the pool. But all his family left and he couldn't go with them. We took him to a vet but the vet said he wouldn't survive and we had to put him down. It was terrible . . .'

Judy is Australian. She met her husband, Denis, an Irishman when they were young, in Rome. They have lived in Madrid since the 1970s; their house has long been a hub for visiting poets and playwrights, Denis having organised the Irish contribution to Expo world fairs. An actor, with a vast repertoire of one-man shows, he has become one of the most prolific and successful theatre directors in Spain. He is a quiet man, bearded and owl-eyed like a Shakespearean player.

'Because I'm not Spanish, you know, they don't ever review me,' he said, sadly. 'They review the actors all right, and at the bottom they just write "Directed by Denis Rafter".'

He was rehearsing *The Merchant of Venice* and invited me to meet his cast and watch them work.

'Believe me, sir, had I such ventures forth,
The better part of my affections would
Be with my hopes abroad. I would be still
Plucking the grass to know where sits the wind;
Peering in maps for ports, and piers, and roads;
And every object that might make me fear
Misfortune to my ventures, out of doubt
Would make me sad . . .'

This is Solanio, a friend of Antonio, diagnosing the causes of Antonio's mysterious melancholy. Though Antonio denies it, Solanio surely has it right. Antonio has invested heavily in trading ships: his forture is all at sea. 'That I have much ado to know myself', Antonio ponders at the opening of the play, wondering at his sadness. I felt the journey had brought me closer to mystery, both beautiful and awful, than I had ever been. I felt as though, in casting everything off, I had lost the world but gained something of myself. My love of friends, and my work, and this strange, sudden, ignorant yet complete thing I felt for someone I did not know – these things could not be cast off and did not waver.

Shakespeare gives swallows a line in *Antony and Cleopatra*. His source is Plutarch, who reports that swallows nested under the prow of Cleopatra's flagship before the battle of Actium. Plutarch writes that this was thought to be an omen of the disaster that followed. In ornithological terms it is certain that swallows would not build on a battleship that was much used: the implication is that Cleopatra's flagship spent a lot of time tied up or anchored in port, which would suggest that neither it nor its crew were battle-ready. Shakespeare transplants the incident from Actium to the third and last of the three confrontations between Antony and Octavian.

SCARUS   Swallows have built
          In Cleopatra's sails their nests. The augurs
          Say they know not, they cannot tell, look grimly
          And dare not speak their knowledge.

Augury was the Roman practice of studying the flights, habits, doings and entrails of birds as a means of foretelling the future. Our word 'auspicious' derives from this ancient method of divination. We still use augury, albeit in a very limited, almost unconscious way, part superstition, part empiricism. Children still recite 'One for sorrow, two for joy . . .' at the sight of magpies. The call of the first cuckoo invariably inspires someone to write to the newspapers with news that summer has arrived. Birds flying high herald good weather, gulls inland mean storms at sea, crows and ravens are still birds of ill-omen, the chinking of blackbirds signals the approach of evening, the arrivals of fieldfares, redwings and skeins of geese are sure signs of the coming of winter. Of all birds, though, the swallow carries perhaps the greatest weight of prophetic folklore. Long before the battle of Actium, Greek sooth-sayers saw the future in the behaviour of the birds. On the eve of his departure to fight a battle against the Medes in 334 BC, Antiochus, son of Pyrrhus, found a swallow had built a nest in his tent. It was believed to herald disaster, and indeed Antiochus lost the fight.

So much human effort, certainly the effort of the culture from which I come, is directed, and has been since the Enlightenment, into demystifying and explaining this phenomenon, the question of guidance, of fate; the question to which the answer would be the meaning of life itself. Once the answer was simple, or at least the region within which it lay was clear – religion. Since the fall of God from His primacy over western thought, science has set about unlacing the obscurity which surrounds what Hamlet calls the 'divinity that shapes our ends' by looking ever more closely at us: at our genes, our psychiatry. The idea that the determination of our fates lies outside us, outside our bodies, our histories and experience, remains the province of the devout. What happened to me as I followed the swallows across the boundary between two worlds, from Africa, where unseen and invisible powers beyond science are alive and central to many lives, to Europe, where unseen hands are always assumed to be human or technological, and all explanations must be rational, suggested to me that a belief in either system which dismisses the other is a misplaced and limited faith. Perhaps something similar would have happened whichever route I had taken; though there is no telling, I am certain of two things: first, the Zulus, in my case, were right. I did follow the swallows and in a way I did not come back. And I did find what I was really searching for. It may sound strange, but having chased them to the furthest south, then followed them north, travelling as they did, sometimes singly, sometimes in loose groups, and loitering in some places, as they did, plunging through and over others, it seemed to me a very natural miracle that in following them I should have come across one of my own kind.

'I only really feel at home when I am a stranger,' Rebecca had said: my feelings exactly.

The swallows and the girl were now the better part of my hopes abroad. To follow and find them I owed the authorities their pound of flesh.

There is an intriguing diagram on the wall of the British Consulate in Madrid, produced by something called the Identity and Passport Service. It is a graph, on the horizontal axis of which is 'Better security', and on the vertical axis 'Better technology'. The plot shows how British citizens have identified themselves over time, beginning with a passport, passing through computer readable passports and progressing to biometric passports and identity cards carrying chips bearing data unique to the holder. At the top of the plot, at the ends of the two scales, the diagram dissolves into nothingness, as if the designers have stopped short of the full implication of their model which is that complete security will only finally exist when perfected technology allows the merging of the citizen and the passport – when we become our passports, when we carry the chips implanted in us. Waiting in front of an armoured glass screen, taking turns to approach the counter, was a scattering of individuals, couples, single women, lone travellers and students who all fell woefully short of this ideal. We had no chips or passports; we were not computer-readable. It took two visits over two days, photographs, forms and money but remarkably little fuss, and I was issued with a sheet of paper, signed and stamped with a picture affixed, which proclaimed me to be a British citizen travelling on an emergency passport, issued in Madrid and valid for five days. Five days was the maximum allowed, an official explained. He would have preferred to issue it for twenty-four hours, to permit me to take a flight the next day, but if I insisted on travelling overland this was as much leeway as he could allow. I must be back in Britain by 19 April, he said; if I missed that date I would be in trouble – did I understand? Touchingly, since the armoured glass prevented us shaking hands, he took two of my fingers through the little slit by which documents could be exchanged and shook them.

'Good luck,' he said.

I took a train north to Zaragoza, on the Ebro River, a swallow highway in the autumn, and, I calculated, somewhere they were sure to be found in the spring. Spanish trains are as thoroughly protected as aeroplanes: boarding one in Atocha Station you cannot forget the massacre of commuters that took place here one morning in 2004. You pass through a security screen and X-ray machines. You are required to carry and produce your passport.

Zaragoza was half a building site. African men were labouring on the roads: there were fewer of them and they had more machinery than their equivalents in Congo, but otherwise it was almost the same scene. Teams of northern Europeans were arriving at the station to supervise and complete the works, all in time for the Expo 2008. I walked into town; it was a cold bright evening and a well-to-do city; Zaragozans were taking what looked like their first evening promenade of the year, all done up in smart coats. Among them I fell into conversation with a young man from Senegal who was walking quickly, and shivering. He was thin and quick-eyed, he moved as if trying to take up as little space as possible and dropped his gaze when we passed townspeople; their glances slid rapidly off him. He was working with a friend as a decorator, he said. He could not have been more than twenty-five.

'Spain is hard,' he said. 'They don't like you here and it is incredibly expensive.'

'Where do you live?'

'With my friend, there are four of us there.'

'And what is your plan?'

'Save money if I can, until I can go north.'

'Where do you want to go?'

'France – or UK.'

'Why?'

'I have heard it is better . . .'

'How did you get into Spain?'

He smiled vaguely, shook his head and would not answer.

The Ebro was a yellowy green which you shivered to look at; it ran hard and cold with snow-melt from the Pyrenees. The old town of Zaragoza lines its south bank and clusters around the cathedral, the Basilica de Nuestra Senora del Pilar, a heavily domed, turreted and dim-hearted thing built around a marble pillar on which the Virgin is supposed to have appeared in AD 40 in the middle of a sermon by St James – Santiago. Pilgrims come in busloads to kiss a portion of this pillar. Ignored, by comparison, is a miraculous building on the south side of town, near the bus station – the Aljaferia, once a Muslim palace, complete with a courtyard of interlocking arches and the remains of a mosque of stunning beauty, all incorporated, in the fifteenth century, into a palace for Fernando and Isabella. Despite being as beautiful, in miniature, as the Alhambra of Granada and one of the pinnacles of Hispano-Muslim art, the Aljaferia is almost bereft of visitors. The Catholic and Aragonese parts of the building are heavily signposted and supervised; the Islamic parts relatively empty. It is as though there is no room in modern, northern Spain for certain parts of its history, while others are preciously preserved. In the basilica two bombs hang from a pillar: they were dropped on the structure by the Republicans in the civil war, but failed to explode.

I crossed the river as the evening came on, and there were swallows swooping under the bridge. Following a walkway along the river bank I came on a great commotion of police and ambulances: they were pulling a body out of the icy water.

My Spanish was now much better; by beginning conversations with an apology for not speaking Spanish I found I was able to elicit help and understanding: this produced a cheap meal and a cheap room.

Zaragoza went to work with a cold morning but a warm sun. By the Ebro, not far from the cathedral, I found an old lady leaning on a wall and staring down at a patch of sand on the river's edge. She was watching swallows. It was the first time since Bloemfontein I had found anyone taking notice of the birds, and the first time I had seen them on the ground. They were drinking and dust-bathing, ruffling their feathers and twittering. They looked strange, out of their natural element, the air, little boat-shapes, like beached yachts on stands. The old lady and I did not understand each other, except for one word.

'*Golodrinas! Golondrinas!*' she kept saying, excitedly, and pointing. In Spanish mythology the *golondrina*, the swallow, is supposed to have tried to remove the thorns from Christ's head as he hung on the cross and in so doing pricked itself, which explains the red patch on its throat and face. The same red patches are explained in parts of Russia by a story of swallows trying to remove the nails pinning Jesus to the cross, while in Sweden the bird is said to have sung a song of consolation to the dying man, and its name in Swedish, *svala*, also means 'to console'. But it was surely the consolation of the turning seasons that excited the old woman. She pointed at the sun, and then at the birds, and smiled and chattered delightedly.

I took a bus eastwards, paralleling the Ebro, to Barcelona. The comfort and cleanliness of the thing, its punctuality and space – one person one seat! – all seemed an amazing luxury, but my neighbour was miserable. He wore thin clothes, shrank in his seat and seemed to be hiding from the world beyond the window by manipulating the curtain. I asked him what was wrong. He was from Ivory Coast, he

said. We were both delighted to find someone else who spoke French.
He was going to Barcelona to stay with a friend and look for work and
then he hoped to go north – it was the same story again. How had he
got into Spain? He had flown from Morocco. Why was he so nervous,
then? He said he was scared of the Spanish police – his friend had
warned him to stay away from them. I believe his visa had expired.

'The problem is,' he said, 'I do not have any money to reach my
friend. He is outside the city, in Badalona . . .'

I gave him 20 euros, which he stared at in disbelief, before kissing
the note and crossing himself. He took my hand and hung onto it,
pouring out thanks and bad breath.

In the seat behind me was Vicente, a young Aragonese with a thin
mournful face and long dark hair tied in a pony-tail. Vicente lives with
his parents in Zaragoza and commutes to Barcelona to study composition – he is a guitarist – and to stay with his girlfriend. When I
explained what I was doing he said, 'Ah, so you are travelling in a
permanent spring.'

'Yes – it does feel like that.'

'These are the first bright days we have had,' he said, as the coach
drove eastward through biscuit-coloured hills.

'Here in Aragon we say the swallow is the bird which melts the
snow.'

He taught me a rhyme: '*Hasta el cuarenta de Mayo, no te quites el
sayo*', which means 'Don't take off your coat until the fortieth of May',
a reference to the way spring warmth drags its feet in the cool shadow
of the Pyrenees. We passed a sign marking the Greenwich meridian.
Ever since Morocco I had been confused about time: Morocco follows
Greenwich Mean Time, putting it one hour behind continental
Europe, despite the fact that most of the country lies east of the longitude of Lisbon. Here we were, crossing the meridian, theoretically
moving forward an hour, though neither time nor the clocks
distinguished between one side of the line and the other. Like birds,
we take our cues from the seasons, from the phases of the moon and
the movements of the sun. But we have formalised our calculations
into a rigid but invisible web of grids, of time and of space, which

theoretically tell us when and where we are. The problem is that though there are many repeating mathematical patterns in nature and cosmology, the rhythms of the earth fluctuate outside the calculations we have designed to contain it. A September day in northern Europe may well be hotter than one in July. We talk of early springs and late summers as though the seasons were somehow out of joint, while it would perhaps be more logical to consider that it is our neat calendar of hours, days and weeks, with their chain of 'seasonal' festivals, that is inaccurate. The attempt to unify human and natural time through a melding of Christianity and older, pagan myths is a common theme in the mythology of swallows. In many European countries it was believed that swallows arrived on the 25th of March, flying down from heaven on the feast of the Annunciation, bringing warmth to the world, a tradition that was particularly strong in southern Germany. Swallows were supposed to arrive on St George's Day in Mecklenburg (23 April), on Palm Sunday in Saxony, on the feast of St Gregory in Bergamo, Italy (12 March), or if not, on St Joseph's Day (19 March) or two days later, on the 21st of March, St Benedict's Day.

Travelling in a permanent spring, as Vicente put it, made me question the assumptions of orientation on which I had built my conception of the world. North, I learned in geography, is 0° on the compass, up the page. But these are mere impositions, commonly agreed, like the imposition of human time on time itself, which allow us to agree on the orientation of space. Swallows do no more fly up and down the world than do they pay heed to days of the week. Indeed, in the rhythm of their seasons, it makes more sense to imagine them flying on a line of latitude, or east–west, than it does to say they fly north–south. Born in high summer, they fly in late European summer towards early African summer, then repeat the trick in the other direction. In relation to the movement of the earth relative to the sun, their migration questions our very notions of stasis and travelling. In the terms of the universe, of space beyond our planet, swallows maintain a much more consistent distance from the sun than do we who stay at home. While the world turns under them, their vast journeys hold them much closer to a single point in space than the 'fixed' barns and outbuildings they nest in, which are conveyed thousands of miles to and fro by the elliptical motion of the earth around the sun, and the 'wobble' of the planet on its axis.

CHAPTER 11

South to North: Barcelona
to Calais

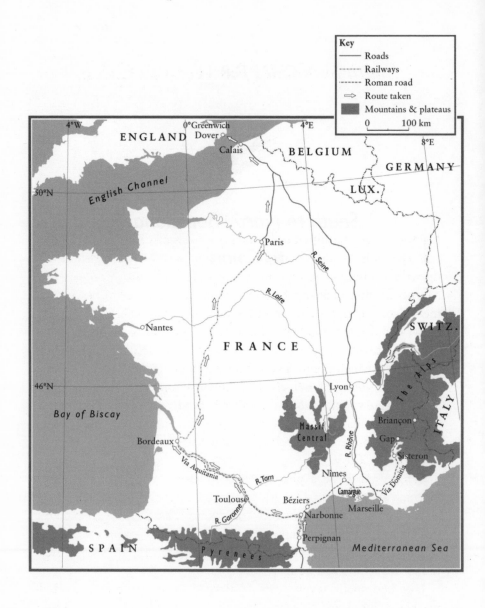

South to North: Barcelona to Calais

There were tall blonde Lithuanians behind bars, Gambian boys selling hash and cocaine on the waterfront, French girls loading up with alcohol in the supermarkets, Slovenes in the internet cafés, Moroccan and Brazilian men fighting outside a club on the beach, and men, women and their children of every nationality in Europe eddying down the Rambla and circling in and out of the little streets of the old town. Above the statue of Cristobal Colomba, Christopher Columbus, which points a commanding finger south-east to the sea, urging the young of Barcelona to go forth and seek their fortunes abroad, were swifts, gulls and parakeets. I only saw one swallow in the city: in the Plaza George Orwell, where the city governors, in a sardonic tribute, have erected a sign warning that the square is covered by CCTV.

'This used to be a place for drugs,' said Jake, a wryly smiling German architect I had known a little in London. 'Lots of acid.'

We looked for the cameras but saw none. There were a great many dealers by the port and on the Rambla. In Africa they would have been selling SIM cards, but entry level into the Spanish economy seemed to involve selling bad hash and cut cocaine.

Jake had spent seven years in Barcelona, working in a glutted market. 'There are so many architects here,' he groaned. 'They come from all over Europe – so they work for nothing.' He had landed a job in London and was making preparations to leave, while

lamenting Barcelona football club's decision to sell their best
players.

'But if I had to get a mortgage perhaps I would get it here. It's
pleasant . . .'

It looked delightful to me, as the lights came up around the port.
'It's perfect!' I cried. 'What's wrong with it?'

'The tourists,' Jake grimaced, mourning the masses like the subject
of an occupying power. We watched a line of them disgorging from
the airport bus. 'See, the Ryanair People,' he said, grinning, 'your
beautiful countrymen.'

Pallid, overweight, and over-burdened with baggage, they did look
resoundingly British. Each one trailed a bag on wheels. The adoption
of these devices means it is now universal practice among the peoples
of the north and west to travel with more luggage than we can carry,
as if we have become so attached to our possessions that we must take
as many as possible with us, unable to contemplate existence without
them, even for a holiday, like children who cannot leave home without
a sack of toys.

I sat there, smoking and smirking in an unattractively superior
manner: I did not reflect that at least the travellers would not be mad
or stupid enough to throw their suitcases into the sea; I conveniently
forgot the weight of my own rucksack, when I began, and it never
occurred to me to catch myself in that most Brit-abroad of snobberies:
labelling fellow-countryment tourists, in contrast to oneself, who is,
of course, a traveller.

'And the regional thing,' Jake went on. 'They are so mad about
identity here. There are no theatre productions in Spanish in
Barcelona, only in Catalan. It makes it very provincial.'

We ate in Barceloneta, a fragment of the city near the port which is
still a fisherman's and sailor's quarter, then drank Dominican rum at
the coach station, waiting for my bus to Perpignan. We said goodbye
and I boarded. The bus was wild.

Almost all the seats were taken, except at the back, which was the
domain of a band of Kosovans. There were three men and four
women, all commanded by the youngest of their party, a girl in her

early twenties, pale-skinned, dark-haired, quick and full of a ferocious aggression. I had barely sat down when she demanded money. I refused, so she demanded water. She returned the bottle contemptuously and, deciding that she wanted more space than the single seat she occupied, opened a mobile phone and made it play music, loudly, until her neighbour complained. She stared him out and replayed the music until he got up, cursing, and went to seek another seat. Then she curled up like a cat and apparently fell instantly asleep.

We crossed the frontier into a country I had thought I knew, but which in the light of what I had seen of its former empire I now saw afresh. A country which consistently tops polls of where Europeans would chose to emigrate to; the country which gave us the terms left and right, in which we still think of politics; the country which more than any other held the key to the continent's direction after the Second World War; the country which has given the West its most radical and influential philosophies since the Enlightenment, from existentialism to deconstruction; the race whose sensibility and language is wonderfully revealed by terms which English cannot translate: *savoir-faire, laissez-faire, hauteur, demi-monde, bon appétit* . . . a language which constantly forces you to evaluate and avow your relationship with every interlocutor – *tu* or *vous*? A place in which all understand the invisible bounds which define a citizen's place: I will never forget being frozen in my teenage tracks by the blistering distaste with which an elderly Parisienne addressed me: '*Monsieur, vous n'avez pas le droit.*' ('You do not have *the right.*') I had not realised you were not allowed to walk on the grass.

On the French side of the Pyrenees, where migrating swallows used to be trapped in nets for food, they have an expression, '*Tu as les yeux en couilles d'hirondelle*': you have eyes like a swallow's balls. This refers to dark shadows under your eyes, contrasting with a pale face: like the contrast with the white under-parts of a swallow, where it meets the dark of the tail. I think my eyes may have looked something like these *couilles*, at Perpignan Station at four in the morning.

The town was abandoned to silence, except for a spill of light from the Hotel Terminus. The bar was officially closed, but they let me in: it was like walking onto the set of a play. Under a single dim bulb four figures were gathered around the bar. Bertrand, twice divorced, a ruddy-faced roue and rogue held forth to Christophe, Christophe's girlfriend Hélène, and Xavier, the put-upon barman. Because they were 'closed' we could smoke.

'The best stories are about love!' Bertrand cried, as we all hit the Dominican rum. He and Christophe discussed the relative merits of women as Hélène drank and giggled and nodded along, adoringly attentive to everything Christophe said: old women versus young women, foreign women versus French, Parisians versus the English – 'Damn the Parisians and the Foreigners!' cried Bertrand, slapping me on the back. They fantasised about going upstairs and seducing the two Colombian girls who had taken a room and who were not that pretty but would do, it was decided. Xavier should go and wake them and bring them down. Xavier demurred, was roundly abused, and more drinks were ordered.

'Women are mad – they must be – two of them married me!' Bertrand exclaimed.

The town woke up quickly at dawn. Shutters went up, bread vans did the rounds, we were all shooed out of the bar while the new barman put the lights on, pulled the chairs off the tables and opened the doors to let the smoke out. Then we were allowed back in, for croissants and coffee.

The first customer is a little old man in a blue blazer. He has a thick southern accent and his voice is a growl as he hails me.

'*Arrnglais?*'

'*Oui – Gallois, alors, mais oui – Britannique.*'

'*. . . ils sont coquilleurs . . .*' he said, with disgusted scorn, flipping his hand over and back, '*coquilleurs . . .*' flip, flip. His name is Yves. I am intrigued by this antagonism and delighted by him, so happy to be able to speak, here, after Spain, where I was reduced to infancy by ignorance: Bertrand and Christophe kept using words just beyond me. They would explain, translating their argot into my French, but not Yves. He mimes. But why does he hate the English so much? What does it mean, '*coquilleurs*'? The gesture seems clear enough.

'You can't have confidence in them?' I hazard.

'*Coquilleurs* . . .' He grimaces.

'Listen,' I say, with appropriate heat, 'in my country – Wales – we have a real history with them. What is your history?'

He touches the lower lid of one eye, pulls it down.

'*La guerre.*'

'The war? We – they – helped you in the war . . .' ('*Nous avons – ils vous ont fait une service . . .*')

Yves is seventy-two. 'Four when the Germans came, eight when they left.'

In 1957 he was in Algeria.

'You were a para?' I exclaim, amazed. The barman, who is listening closely, snorts and laughs. Yves shakes his head. He was a Marine; he fought in the hills.

With something like relief we get into it, raising our voices, seizing the other by the arm, underlining our points with grips. He is very small and our debate makes me feel almost as vivid as him, despite the long night.

'Oran!' he cries. 'Mes-el-Kebir!' (Where the Royal Navy sank half the French fleet rather than risk its falling into German hands.)

'There were sailors in the ships . . . *Et les Arrnglais . . .*'

He makes a horrible thrusting movement, calculated but forceful and with a twist, like a man picking out the killing spot of an opponent, stabbing between the second and third rib.

'The British Empire!'

We start to make a list, I start to write 'our' countries names in two columns: we agree there are too many. Africa splinters, east to the British and west to France: there at least we were equal.

'*Le vieux faucon,*' the barman calls him, proudly – the old falcon – with another affectionate laugh.

'Do you think it was a good thing, Yves, in the end, colonialism?'

He makes me repeat the question. Tears come to both his eyes. He did thirty-five years for SNCF after Algeria: hence his regular place at the bar of the Hotel Terminus at seven-thirty, his huge popularity, his – something like – totemic status among all the few who are around now, opening up for the day, for this day, 17 April 2008.

He has a divorce, a daughter of forty, a house in Perpignan, and he shows me a brown and white postcard of his birthplace, Cerbere, in 1915, looking like heaven (like an Algerian heaven, actually): a little white fishing commune on a bare brown hill by the sea. It's changed, now, he says.

Swallows mean good weather, he says.

'I'm talking about Les Arrnglais,' he keeps saying, as if to reassure me, as if to implore, to save me from taking it personally, which I am not, and am.

'But do you think – ?'

I kept coming back to the question. The tears which fill his eyes are so fine they will not run. Now he is imploring.

'There are people there who – wish we had stayed – want us to go back – without us – Africa – would not be developed . . .'

'If you have money,' he said, 'hide it'.

That day was a confusion of sleep and sleeplessness in which I travelled hundreds of luxurious miles courtesy of SNCF: the cleanest, quickest, most comfortable and reliable system of transportation in 6,000 miles, a marvel of French style and efficiency and a huge gift to European travel. French passengers seemed remarkably displeased with it. Queuing for one of the wood and steel ticket desks, with its little lamp and sleek machinery – a world away from the grimy windows and shouted exchanges of the British equivalent, more like an appointment with a sympathetic travel agent than a ticketing system – I was amazed by the displeasure and frustration of others in the line. The man in front of me complained angrily about the timetable and the prices.

'Well I think SNCF is wonderful,' I informed the clerk, drowsily.

'So do I!' she exclaimed. 'But people will always complain.'

I arrived in Narbonne in a warm rain, in time for the morning market. Narbonne is a crossroads of canals and ancient footpaths. The Roman Via Domitia which linked the Alps to the Pyrenees runs through the middle of town. A section has been uncovered: you place your feet where once Roman sandals trod. In the market a man in young middle age with a mass of curly hair was setting out a second-hand bookstall.

'I am Parisian, originally,' he said, 'but I gave it up for the South because here is the art of life.'

A cheerful girl in a bakery and coffee shop played the lead in a series of sketches of a dead-pan polish that could only come from daily rehearsal.

'Good day, Madame!'

'Well, it's a day, anyway . . .'

'Good day, Monsieur!'

'Good day, Miss.'

'What would you like?'

'Another day. And some bread.'

'Good day, Madame.'

'Well, we do our best . . .'

'Good day, Monsieur.' (This to a policeman.) 'How is it going?'

'It's going, it's going, but it's not gone yet . . .'

They discussed something. 'Well that's life!' she concluded.

'It's *a* life,' corrected the policeman.

I sipped *café au lait* and thought of Eric Newby's description of the French:

A people with the oldest colonial tradition in Europe who built up one empire, lost most of it and then started all over again to build another, the second biggest in the world, in which no one in what the empire builders called *La France de la Métropole* was ever very interested. Voltaire's description of Canada as 'a few square miles of snow' was not only witty but typical of what *Les Métropolitains* felt about it . . .

In that bakery you could scarcely feel further from Brazzaville, from Douala, from Birni n'Konni, from Casablanca, even from Algiers, and yet something of the same scene is played out in each of them, over the same food and drink, in the same language, and therefore through something of the same sensibility every morning of the week.

On another train I zig-zagged up to Toulouse, where the station café was full of Algerians and Moroccans, young men, their skins pale and ill-looking through lack of sun, smoking dope and flirting with the waitress. One of them led me to an internet café.

'I am Moroccan, yes, but I am also French – Toulousian. My father came here but I was born here. It's nothing special, here. It's the Birmingham of France!'

'Are you going to leave?'

'And go where? This is home.'

The next train followed the track of the Roman Via Aquitania, curving north-west up towards Bordeaux. It was a beautiful spring day here, perhaps the first of the year in Aquitaine. Ten miles west of Agen a crowd of swallows swooped over the river – here the Tarn joins the Garonne, draining the whole of south-west France into the Gironde estuary and the Atlantic at Bordeaux. On the north bank the land was a dark green; a damp, bright shade, the first time I had seen the colour of home since January.

Tiny gradations of changing soil and vegetation bleed earth's colours into one another almost imperceptibly: you do not notice the sandy yellows of the veldt becoming the pastel ochres of Zambia becoming the dark gut-red of Cameroon, unless perhaps you over-fly them all in less than a month. Perhaps to the swallows the earth is a colour-coded route map of chromatic change.

The train passed fortified *châteaux* and the sites of much older forts, castles on peaks: to travel this way would have been like journeying in Congo, once; after the Romans left and France fractured into Mérovingien and Frankish principalities, soldiers extracted border taxes on behalf of each fiefdom. On the south side of the river the land was fairer, with blonde grasses. Bright blossom like dollops of ice cream covered the orchards and there were palm trees in the gardens, as though the Via Aquitania divided northern from southern Europe.

In the evening, in the last of the sun, we came into Bordeaux St Jean, a station which seems to have fallen out of a Monet: the roof is a fretwork, functional, no doubt, but which multiplies into a cobweb of thin dark iron below the stained glass. The city presents an impeccably neat, grandly high-bourgeois façade, its history of wealth displayed as space, diamond-shaped squares aligned with the river. Bordeaux seems as needless and heedless of the rest of the country as France itself is of the rest of the world: to get out of here, you must take the TGV. It is a spring evening but the tail of a long winter still twitches in the air; the wind whips cold off the wide Garonne and there is no heat in the sinking sun.

'The old Bordeaux people are very snobby, very bourgeois, very posh and very mean,' says Fleur, lightly: she is a student of International Relations and the receptionist at a hotel near the centre which is too expensive for me, as are all the places I try. This trading city is far beyond my means.

The train for Paris leaves at 04.40 but the ticket includes a blessing: the right to the waiting room, where men and women, many of them black, curl up on their luggage and sleep. I wake with a start, having set no alarm, and am out, hurrying through the underpass, up onto the platform and into the train before I really know what time it is, what train I am boarding or whether I am dreaming. The doors shut behind me and the train rolled. My inbuilt alarm clock never cut it so fine, before or since.

TGV 8406, the 04.40 service from Bordeaux St Jean to Paris Montparnasse is under the control of a tall, saturnine conductor who would inevitably be portrayed by Gérard Depardieu, should it ever be the subject of a film, a man with black crocodile-effect cowboy boots protruding below the turn-ups of his grey SNCF uniform, who despatches the train at every station, after he has had just as much of his cigarette as he wants, according to the sweeping gold hand of his Breitling aviator watch – the same watch Sarkozy wears, he informs me. This train does not leave to the minute, but on the second. We meet in the buffet, where he considers me, in between flirting with the girl behind the counter who makes us coffee.

'English? I hate the English,' he says, without a flicker of a smile.

'Why?'

'You are destroying Europe, destroying the world, destroying everything. You and your liberalism. The first Gulf War, Lebanon, Palestine . . .'

'What are you listening to on the i-Pod?'

'Amy Winehouse!'

'You like her? She's English . . .'

'She is wonderful! And Massive Attack – they are *formidable*.'

He has written a novel. 'But it is not commercial,' he says with a shrug. I half-expect him to blame the English for this.

'What's it about?'

'It is philosophy. It is about a man . . .'

The man sounds remarkably like him. I am not clear on the philosophy of the philosophy but it is a very big book.

'Too big, I think,' he sniffs. 'I sent it to a publisher but . . .' He shrugs.

'And what is the problem with the English?'

'The Anglo-Saxon way. Sarkozy is going to privatise SNCF, you know, they are going to destroy it.'

'Why?'

'Because they are jealous. That is the Anglo-Saxon system. Thatcher, Blair – all the same. Like the Americans . . .'

And Sarkozy?

'*Une petite mouche*,' he snorts. A little fly.

The spring dawn passes in a beautiful dark-blue blur, lightening as we go north through wooded hills and farmland. There are armed police on the train and soldiers with rifles at the station. Guns have been ubiquitous all the way from Cape Town, only the makes have changed.

We made Paris just before rush hour. At the top of the platform I turned to admire the engine of the silver TGV, half-missile, half-serpent. Plastered over half the windscreen in a crimson slur was a large, violent bloodstain and a single feather.

It is a long and delightful walk from Montparnasse east along St Germain heading for the Seine, Charonne, the Porte de Montreuil and the bus station. Spring is coming to the boulevards, sprays of buds like flicked gold paint, leaves brighter, paler than the green of Aquitaine, and then there are the chestnut trees, holding their little white candles bashfully against their new green skirts, like debutantes. A procession comes up St Germain, eighty chanting students escorted by police. They are holding placards and shouting 'No!' to tuition fees. '*Révolte!*' They look very young and laugh self-consciously; the police guarding them smile indulgently and passers-by nod approval: the next generation of the bourgeoisie doing as they should, cutting their revolutionary teeth. In a shop specialising in prints and maps of North Africa the proprietor, a Tunisian-turned-Parisian who has been here for twenty years, says he hates the English, 'the dogs of the Americans': 'Which is worse,' he asks rhetorically, 'the illness or the whore who gives it to you?'

He remembers swallows migrating through the Tunisia of his childhood. 'We used to shoot at them with our catapults,' he remembers, smiling fondly. 'I got one once! It wasn't easy . . .'

The route bends past Austerlitz and the Gare de Lyon, along the Boulevard Diderot, over Nation, along the rue de Pyrenees where workmen are planting a lovely young Japonica tree outside the Bar Biarritz, and through old Charonne, once a separate town, with a handsome church and a village-like square. At one intersection a plaque records that a salamander stopped here once, and rested.

The bus station is a slow world. People move slowly, work things out slowly: we are the time-rich, cash-poor of Europe.

An Algerian man is waiting for a friend on the delayed bus from Madrid.

'Yes, I live in Paris now. I hate it. It's shit, I'm going to move to London. How much is it to go there?'

'Forty-five euros, plus five in tax.' (The first departure tax anyone has asked in the entire journey.) Because he was the first Francophone not to proclaim a hatred of the English I told him he was welcome to Britain, and advised him to try not to arrive in the winter.

The motorway from Paris to Calais is full of trans-continental traffic. According to its markings one truck belongs to a Russia–Poland–Spain–Morocco transport company. A British vehicle belonging to C. J. Bird Transport is decorated with a huge swallow. The lorries cross the wide flat spaces of the Picardie plains like tramp steamers on a dead sea. A sign proclaims 'Picardie – Terre Fertile' but the earth's fertility is being flogged remorselessly: sprayed, ploughed and planted. There are no hedges or fences; gangs of tractors permit the earth no rest. This is a land which has been made addicted to chemicals; nothing lives here but a scattering of crows. There is nothing for a swallow to eat and no sign of them. Outside the old mining villages are spectacular slag heaps, straight out of a Zola novel, and spring covers the distant hillsides with buds. The country reminds me of Gulliver in Lilliput: mighty, in its sweep to the horizon, but tied down by the million tiny ministrations of men and machinery, enslaved.

The coach stops on a windy stretch of tarmac between French and British customs. Unclear on which way they should go first, the fifteen travellers fragment, some heading one way and some the other. The customs officers swiftly realise that half the head count provided by the driver is missing and immediately, comically, blame each other. The English are liars, mutter the French officers, darkly; the French are playing silly buggers again, say the English. An English officer is on the phone to his boss.

'Yeah, yes, I know I said I was going to do that thing for you but I haven't even had time to start. As it is I've only got eleven of us on. Eleven staff and one Immigration Officer for the whole port of Calais . . . It's a complete . . . That's all thanks to the new – you know. It's UK plc or whoever . . .'

They stamp us all through and the bus drives us onto the ferry. The boat seems a wonderfully old-fashioned cross-Channel world:

children wailing, adults nattering, the first British voices and the first Welsh, too. An excitable woman on the stairs informed her family 'So now we know what's up there we wanna see what's down here!' and they follow her, complaining contentedly. It was a dizzying, elating experience to be on the ship, amid the throng, to look around and to see myself no longer as singular, but as one of a crowd, of a type – of a people. All the way from Cape Town, I marvelled, and here I am. Almost home! The matter-of-factness of the British seemed to preclude any great celebration. Cape Town, I could imagine them saying, what a long way; was the weather nice?

Before we cast off the captain's voice addresses us in calm, estuarine tones reminiscent of an Essex golf club: 'Well, ladies and gentlemen, we're ready for sea now; conditions in the Channel . . . there was a gale earlier on but now it's overcast . . . murky . . . and with our stabilisers we should be fine, it's just a stiff wind from the east. Sailing time is about one and a half hours. I do hope you have a pleasant crossing.'

'I hate boats. Make me sick!' announces an elderly lady. 'That's why I'm out here.'

I see her later, mid-Channel, wrapped up on the veranda deck under the misty sun, as we steam through a calm eye in the veering wind. She looks completely happy.

A lorry driver from Manchester is leaning over the rail, eyeing the listless grey surging of the Channel. 'You should see it when it's rough or blocked because of strikes or whatever. Operation Stack they call it: anyone can do it, a couple of lorries or a few fishing boats blockading a port – chaos. Normally come over at night – this is the first time I've seen it in daylight for ages.'

His boss made a fortune in VAT washing, he says. 'You take a bunch of mobile phones, drive them to the Czech Republic, take a break while they change the seals on the crates, then drive them back. He made millions. Had a whole fleet of little vans doing it.'

Today he is driving a load of thirty-five pallets of car interiors.

'You sweat it a bit in customs, hoping no one has stuck any tobacco or whatever in there for a mate, because if they have it's you who sits in the cells waiting while they sort out the paper trail . . .'

I spend most of the crossing on deck, shivering happily in opaque sunlight. It would be too much to hope for the sight of a swallow making the crossing; with the wind settled in the north-east they must either be waiting for a shift in its direction, or battling in against it from somewhere further south and west, the Cherbourg peninsula, perhaps, or Saint-Malo.

The cloud breaks and suddenly there are the White Cliffs, and there is Dover. I feel another surge of exultation, and something like trepidation, too. How will I find the world I had left – how will it find me? And though I had faith in them, there was still an illogical question mark in my mind: would the swallows really make it? It seemed extraordinary that after all this way, and having reached the vast farmlands of France, so many swallows should go the extra perilous miles out to these islands in the North Atlantic. Ornithologists have no way of explaining it but that the birds probably began their migrations at the end of the last Ice Age, following food sources, and have continued simply because it is a successful strategy. The urge to return to where they were born is less explicable, and not all do. Young males seem to head for 'home', their parents' nesting area, moving on only if all the spots are taken. But young females are more adventurous. One female, ringed as a nestling in Warwickshire, was found the following year breeding in the Netherlands. She must have met her mate on migration or in their winter quarters.

The white cliffs become greyer as we pass the moles that protect the Port of Dover. The wind is still stiff from the east. People stream down to the car decks as the ferry performs a neat pirouette and glides into its dock, ropes are thrown, the engines fall quiet, the nose lifts and a ramp comes down. The coach drives us over the gap, and we are in England.

CHAPTER 12

A Swallow Summer:

England and Wales

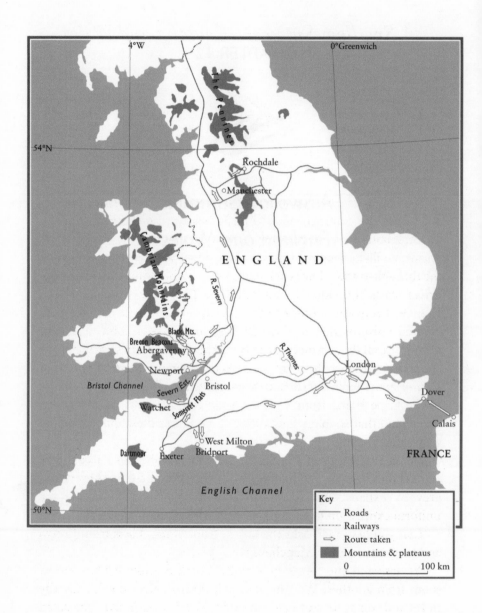

A Swallow Summer: England and Wales

Dover turns a rather weary, almost ramshackle, face to arrivals from the continent. Faded chalk boards offer deals on Sunday lunch; a hotel looks so run-down I long to stay there just to meet its ghosts; an ill-looking man smoking outside a pub gives a blank eye to the through-traffic. The coach hauls us up to the London road and the driver starts the video: a bad American film dubbed defiantly into French. From the motorway there is little to suggest that we have entered a promised land. What Patrice would give to come here, how infinite are all the sacrifices made by all the people from so many of the countries of the earth to get here: France may be the first choice of second home for western Europeans, but from Romania to Kashmir to the Cape Flats, from Turkmenistan to Sri Lanka, it is this low green land that so many long to stand on, under these quietly rolling grey skies.

We enter the city of London at Lewisham. The other passengers stare at the close-packed houses, the traffic, an Italian restaurant dyed grey by exhaust and the multi-national crowds, all wearing the uniform expression of Londoners: part-closed, part-faraway.

'Can you help me?' asks the boy in front of me. He is from Congo and he wants to go to Manchester.

We arrive in one corner of Victoria Coach Station; his bus leaves, soon, from another. We run pell-mell through the crowds, buy the ticket and run again and I am amazed to see that nothing is translated into any language other than English. This is the first monolingual country I have arrived in. The Manchester coach driver welcomes the boy aboard in a friendly way and I am alone again, in a crowd of travellers, simultaneously at home, and lost. I call my father. He sounds welcoming, but cautious.

'*Avoir une hirondelle dans la charpentier,*' the French say – to have a swallow in your woodwork, like bats in the belfry, means 'to be mad'. Did I go mad in Gibraltar? It seems my father fears so. My brother, too.

The 19th April is a bright, sharp burst of spring in the suburb where my father lives. My brother stands on a station platform lecturing me with deep concern. I have lost too much weight. I have undergone one hell of a journey. I am not right. I must, must 'go and see someone' – do I understand? I do understand, but I am not finished. The journey is not complete. We misunderstand each other. I am unwilling or unable to give up my new habits. I sleep lightly and short. I say I am more than happy to go on living without a mobile phone: my father sees paranoia in this. I am amazed at how precious and pressured everyone's time is: did I live like this, once, fitting everything into small windows of freedom between work and rest, like putting chocolates into an advent calendar? I want to talk and talk but I can hear myself talking too much. The English, or Londoners, anyway, never talk too much, except when they have been drinking. I feel I have become both too self-contained and too open for their comfort. The webs of email, text messages, carefully timetabled diaries, pencilling-ins and arrangements that will or will not be firmed up, seem to hold everyone together in semi-presence. The idea that you might not know where you will be sleeping the day after tomorrow, and therefore that nobody else can know, seems rabid eccentricity.

For three days now I have seen no swallows and the familiar discomfort is back. Where are they? I must find them. They will be out there, away from the city, coming in like a tide from the south, spreading over the country. Negotiating the trains and tubes without the deadness of familiarity is very strange; horribly tense, in fact. The line of people on the station platform on Monday morning, holding their breath as the minutes count down to the arrival of the train; the barely restrained rush for seats; the sound of people breathing shallowly; the minutely self-conscious shifting of legs, arms and bodies, shuffling into interlocking positions until the carriage becomes a dense mass of human flesh; the intensifying crowds of travellers and trains as we approach Vauxhall; the cattle-like press down into the underpass; the feeling that a step out of place will cause chaos and collision; and above all, all the time, the never-ending harangue of announcements, admonishments and advice from automated voices, station staff, posters and signs: don't forget this, watch out for that, don't obstruct, closed-circuit television, stand back, allow passengers, unattended items, security, security, security . . . the word is everywhere, like a fire-alarm half-ringing, as though the air-raid sirens are on permanent stand-by, war has been declared, and we await the arrival of bombers. Nobody speaks, smiles or screams but all express themselves in little huffs of disapproval, acquiescence or inconvenience: the silence of so many bodies seems at once sinister and comical. But it is difficult not to admire this multi-coloured, multi-cultural, multi-lingual breed of people, these Londoners, who make their way, every day, in one of the fastest, most expensive, most competitive cities on earth. Tourists look amazed and bemused, as if everyone else is playing a joke on them: they gaze around, mystified, as if waiting for someone to start laughing. When a voice begins speaking behind me I have to crush an urge to turn and join in the conversation. The speaker is a man, and by his accent, a cockney.

'Used to go down there when we was younger. No one 'ad property down there then, now everyone does. And that other place, bit further along. Bognor Regis.'

The small boy with him says something which I miss.

'Mullins? Long as he pays 'is own way, the tight sod. Went to Florida with 'im once an' he came back with the same money he went with. You know those two gold coins he 'ad? Got them off 'is grandmother's eyelids. That's what he's like. Good as gold he is but he can be a right little plum.' (Laughs.)

The small boy asks something.

'Yeah, number two in the world he was, true.'

'He's on Youtube!'

'Yeah, there was one bloke above him, an American. He was a bit good. You're soft on 'im, aintcha? That's because you're as soft as doggy do-do in the sun. Don't worry about Mullins. He's got plenty stashed away. His missus is the most intelligent woman I've ever met in my life and she takes care of 'im . . .'

Oh, the joy of being able to eavesdrop fluently again!

We surface from the tunnels on the conveyor belts of the escalators. There are now digital television screens showing adverts as little films where there used to be defaced posters. Corners and entire quarters of the city have fallen to the wrecking ball and are being remade, but it is the permanence which arrests me. London is grand and mighty: here is the evidence, the acquisition and achievement of empire, from the edifices of the great offices of state to the regal sweep of the terraces, from the cathedral to all the spires, this forest of stone, brass and steel has been culled and collected from across the earth. Cleopatra's Needle seems typical – this ancient obelisk is here a little bauble, a collector's item, towed to the Embankment across seas of time and culture, like a caught fish. The British Museum is a pirate's hoard. Nelson stands on his column at ease, like a landowner surveying his estates. How terribly well London has done. The elegance of the city, its ease with its riches, its imperviousness to time and its indifference to the individual seem to render the crowds ant-like in our scurrying, and termite-sized our little lives, harnessed to serve and maintain it.

I retrieve my old car from under an arch in Battersea and set off west, chasing the sun. It is as though the forward motion of the journey has become a way of life that I cannot surrender. I follow the road south-west to Dorset where I gate-crash an old friend, and leave the next morning, meandering still westwards until at last the road runs down to the coast. I pull the car up beside a small harbour and look around, blinking, to see where I have ended up. There is a low mist on the sea. People are stretching and yawning and turning their faces to the sun, its bright but slender heat magnified by the cold air. Wandering along the quayside I come face to face with the Ancient Mariner, a sinewy, stricken bronze figure, clutching a spar, with the albatross roped around his neck.

'We raised the money for it, the townspeople did,' says the duty curator of the little museum, proudly. The small town is called Watchet and it seems to be a universe entirely of its own, obtusely out of step with time. Two ladies join the curator in a lament for the old harbour, which was replaced with a yacht marina, against their wishes.

'It's only usable for a couple of hours at the top of high tide,' one says, scornfully. 'The rest of the time those fancy boats just sit on mud.' Behind them a picture shows a coastal steamer from Wales eternally half-loaded at the quay, a sepia morning preserved in its spotless photo frame.

'Oh, we used to get the Welsh here, raiding and plundering not so long ago,' says the curator. 'There hasn't been much pillage recently though. The local girls would probably say they miss it!'

On the harbour wall a man is selling organic vegetables from a stall, handing them over with enormous care, as if they were jewels; on a bench a black woman is stretching herself out in the sun like a cat in an over-large coat.

'Lovely day!' she says. 'Like summer.'

'Yes! First day of spring, anyway.'

'Do you think so? You might be right, you might be right . . .'

There is a whistle-blast behind us and a steam train comes puffing along the line just above the town. I look up and see two swallows, flying together, heading up the coast.

I follow them along the coast road. It is difficult not to sing at such a beautiful day; the mist lifts slowly off the channel and though I cannot quite see it, I know home is on the other side. There are more swallows over the water meadows of Somerset, hunting the marshy ground around the river Axe. The road winds down into Bristol, through the Avon gorge, then up to the bridge and the giant, ripped surge of the estuary. Up and over, and we are in a bilingual land again. *Croeso y Gymru*, says a sign, Welcome to Wales.

There are several routes into the hills from the motorway; the one I take goes due north first then branches off, following the river Usk. As soon as you leave the main road you are in a different land: Wales is a small coat made of deep pockets. Suddenly there are little meadows; though it is late afternoon last night's dew has not lifted. Hedge lines are splashed with blackthorn blossom and the woven greens of oak and ash. The sun is slanting down now and fair weather is coming in from the high south-west. Swallows swerve and twist low across pasture near the Usk; their travel still has that purpose about it; they are still coming on. The river is swollen, I see, there has been rain. It runs with a reddish churn. The fields are halfway to recovery after the winter; some still look tired, where the flocks have been, but the colours of others are deeper and brighter. The celandines are out, the swallow flower which gave the birds their Greek name, *chelidon*, because they come at the same time. Our valley seems not to have changed at all. I stop the car, get out and breathe in. There is that smell, the smell of the alders by the stream, the sheep fields, the hay in the barns, and something else, a smell as pale as the sky but as sharp as the earth, which nothing else smells like, only this place. A song stops me dead, rich and twisting, bright notes coming from behind and above me. I turn to look up and there is a thrush in the cherry tree, in full voice, and behind it, just there, over 6,000 miles from where I first saw them, are two swallows, a male and a female. We seem to have arrived at exactly the same time.

Swallows and a thrush come together, famously, in Robert Browning's poem 'Home Thoughts, from Abroad':

> Oh, to be in England
> Now that April's there,
> And whoever wakes in England
> Sees, some morning, unaware,
> That the lowest boughs and the brushwood sheaf
> Round the elm-tree bole are in tiny leaf,

While the chaffinch sings on the orchard bough
In England – now!

And after April, when May follows,
And the whitethroat builds, and all the swallows!
Hark, where my blossomed pear-tree in the hedge
Leans to the field and scatters on the clover
Blossoms and dewdrops – at the bent spray's edge –
That's the wise thrush; he sings each song twice over,
Lest you should think he never could recapture
The first fine careless rapture!
And though the fields look rough with hoary dew,
All will be gay when noontide wakes anew
The buttercups, the little children's dower
– Far brighter than this gaudy melon-flower!

The poet was in Italy when he wrote it, and through his imagination simultaneously in England, receiving sensations from places he knew, knowing them so well that he can be sure exactly what the blossom looks like, and exactly how the thrush sounds. There is the amazement of the alchemist who has actually made gold in his exclamations – the chaffinch is singing on the orchard bough now! – and you wonder if he did not look up from his desk in Italy and see swallows as he was writing: while everything else in the poem is a conjuring, a picturing, the swallows have the power to be in two places at once.

'A pair came today,' my mother says, beaming. 'The 22nd of April,' she says, marking the date on her calendar.

And now there are two pairs, and in the following days one pair begins to nest in the derelict part of the house she calls the back kitchen, and eventually raises two broods.

'The first one came on the 1st of April,' she says, 'and I was terribly worried about it because the weather was awful, snow and rain, and he went away again.'

That was the day, April Fool's, when I asked the girl, laughing but entirely seriously, if she would marry me, and she replied, seriously, I believed, but smiling, that she would, and I gave her my green stone from Zambia, bought from the trusted man and carried surreptitiously across innumerable borders, and she gave me a gold ring, which would only sit comfortably on the third finger of my left hand.

And that was supposed to be that. I thought I would come home, and my Mum would make me a cup of tea, and I would sleep and eat and return to my life, and sit down and tell the story of everything I had seen and done. But it did not turn out like that at all. Within a few days I was in Dorset, teaching a writing course, half bound up in the work and worries of my students and half not there at all, as my gaze followed my mind's eye out of my room in the writing centre to the water meadows and the telegraph wires, as spring stormed through the trees and hedgerows and a large band of swallows swooped over the fields, then gathered on the wires in chittering groups, exactly as if they were planning another journey.

'I see your swallows have turned up,' I said to Noel, a pipe-smoking gentleman who looked like Father Christmas doing the garden, whose wife ran the administrative side of the centre.

'Oh those aren't ours,' he said. 'Those are someone else's. They just stop here to hunt. They'll be on their way north again soon. Ours are still on their way.'

The next day, as predicted, they were gone. I wanted to go with them. I found it impossible to settle anywhere, except where Rebecca was, and soon I found myself in Rochdale, amazed that it was all true: it really had happened, we really did meet in Morocco, and there were *les Deux Princesses* to prove it, and Rosie, and we really did love each other, and I really did plan to spend my life with her, and with her beautiful little boy. I set myself up in their spare room and began work. Sometimes I ached for the lost notebooks, but not often. As if writing it down as it happened had fixed it firmly in my mind, I had no difficulty returning, in my head, to all the places I had been. I can still see it all now, vividly, as if I was just there.

And because I am a romantic, and because perhaps I really had been knocked sideways by the journey, I imagined that what had happened to me would be met with universal rejoicing by my family. It was not. Rather than being happy, they were worried to distraction by me. One hot, still morning in June, the day after my mother's seventieth birthday, I sat in the garden with my family and my Intended, looked up and saw an extraordinary thing.

'My God!' I exclaimed. 'Now I really do know something about swallows which no one else does!'

Half an hour later my brother and mother practically forced me into the car and drove me to see a doctor. I was scared but only half unwilling to go with them. I went partly because Rebecca swore she would not let me be sectioned, and that she would wait and get me out of there when I had seen the doctor, and partly because I was secretly terrified that I might have lost my grip on reality to the point of hallucination.

The doctor asked if I was worried about anything, and I said apart from my family, because they were worried about me, no. He wanted to know if I had had any thoughts of harming myself. No, again. He told me that there was always help available, if I wanted it, and I thanked him, and we said goodbye. But I was worried, terribly worried, because I thought I had seen something that I could not have seen: if I was hallucinating then I was surely in serious trouble.

This is what I saw, that June morning: a female swallow carrying a nestling, her beak clamped behind its head, shooting out of the back kitchen, and away across the orchard, only to return a few minutes later without it. Was it some envisioned metaphor for the way my family seemed to have closed ranks against me, rejecting my restlessness and chatter, ascribing all my plans and schemes and my certainty over the woman I loved to some sort of breakdown?

I kept the incident secret for months, like a flutter in the heart or a pain in the lung that you fear to reveal for terror of its consequences, until I came across this account in *Swallows*, a book by Peter Tate:

One strange piece of behaviour that seemed to be the result of a whole string of mixed reactions was recorded by Mr E. I. Cuthbertson. A pair of swallows had built a nest on a curtain rail in a bedroom at Sedburgh, Yorkshire. On 17 June the first egg was laid, and on the same day it was found broken on the floor some distance from the nest. Another egg was laid on 28 June and then another two. Incubation started, but on 3 July an adult swallow was found dead under the nest. Until 5 July only one bird was seen, but then two birds visited the nest together. One of them began to build up the nest rim despite the efforts of the sitting bird to prevent it, and eventually the nest reached some three inches deep. The three eggs it contained hatched on 15 July. The next morning an adult swallow was seen carrying a nestling out of the window, and a little later another nestling was found on the floor. By the evening of 16 July the nest was empty. This entire episode is completely at variance with the bird's normal behaviour, and a rational explanation is very hard to find . . .

So my exclamation was an exaggeration: I do not know anything about swallows that no one else does, but it was not, thank God, a hallucination.

I struggled with hallucination for weeks: not with visions that I was having, but with the common, shared figments of things not seen but nevertheless understood – the nationwide hallucination of what a society is, what a country is, what it means to be British. I was not the only one. At the village pub people talked in millenarian terms; this thing the radio called 'the credit crunch' seemed to be driving everyone slightly mad.

'You'll have to be here the day we declare independence,' said one of our neighbours. 'If you're not in the valley that day then you're out!'

I looked at the British as though I had never seen them before, as I had all the other peoples I had encountered on the road. It was striking how militarised the country had become. On a train a group of boys discussed their training: the running, the shooting, the hand-grenades. They talked about being posted to Iraq. A middle-aged businessman at their table revealed that he had once been an army engineer. They were quite rough young lads but when they heard this they began to ask him respectful questions, and they called him 'sir'. A few days later, in Wales, I picked up a hitch-hiker: a Royal Marine NCO who specialised in mountain warfare, he said he was convalescing from an injury in Afghanistan.

'What happened?'

'Oh, I fell out of a helicopter.'

He talked about how much he loved the mountains of Wales, and the birds.

'What's your favourite bird then?'

'The goshawk,' he said. 'Definitely.'

'Why?'

'They are the centurions of the bird world,' he said.

I was struck by the doubled nature of the British: it was not malign, like duplicity, but it was as though they concealed half of themselves. They seemed a nation of actors, their outward reason, formality and civility concealing interior lives charged with passion, desire, mystical suspicion and philosophical curiosity. What else could explain the torrential floods of sex, smut, humour, peculiarity and chaos which fill the newspapers and chat shows to which they are so addicted? It is as though each Briton conceals another, our outward forms perpetually playing the straight foil to an inner anarchic comedian. The state we have constructed, however, though doubtless staffed by every shade of individual, and one or two comedians, is nevertheless an embodiment of the former form: cold and formal. While every Briton I met, from London to York, was kind and open to the stranger I felt myself to be, the state was closed and distant.

In May news came of Patrice's visa application: he had been turned down. The immigration officer at the consulate in Cameroon gave reasons for the rejection: he or she could not believe that Mr Kenneth would simply give money to Patrice for his trip to Britain (this kind of philanthropy being beyond the bounds of what the official considered reasonable and unsuspicious); he or she could not see a clear pattern to Patrice's earnings in recent years, and my covering letter, being the print-out of an email I had sent Patrice, was not in an envelope and was not signed with a hand-written signature, and might therefore, the officer concluded, be a fake. If Patrice wished to dispute the judgement he was advised to take his claim to the European Court of Human Rights. The officer signed his or her own message with a number, not a name.

It was a heavy blow to Patrice, and came with a worse one: his father died. In the same period Mr Kenneth was mugged on a business trip to Nigeria, losing a quantity of cash and his passport. I sent money to help pay for the funeral, and Patrice refused to give up. He made contact with some part of the rugby establishment in Italy; when I last heard from him he was on his way north, and had got as far as Morocco.

Another message came from PJ in Congo: after I left the police asked him a great many questions about who I was and what I wanted. He said he was desperate to get out of Makoua and asked for help, but when I asked him what sort of help I could give him he fell silent. Without a passport he is trapped.

For days I walked through the woods and fields around my mother's farm, watching the swallows. Their apparent freedom is an illusion in much of the time they are with us. First there is nest-building, or repair, which means scooping up mud from the edges of the stream, and collecting dead grass, straws and hay. Their nests are like inverted cupolas, wattle-and-daub emplacements on the beams. Then there is the business of mating, and defending their nest-sites and their mates from other birds, and then they feed the young. Thousand of insects must be caught and conveyed to their nestlings' hungry gapes. When the rain fell they rested; the moment it stopped or slackened they were out again, hunting, until last light.

It took me a while to find an equilibrium in Britain. My travels had turned me into a Luddite: I was bemused by the stretches of life that I and everyone I knew had given to telephones, computers and email. My mobile phone boycott lasted about two months: it was amazing how much distress and confusion it caused my family, and how it

exasperated my friends; going without any identity lasted a little
longer. It was liberating, in a way, to have nothing but the photocopy
of an emergency passport with which to prove who I was, but then
Rebecca said if I did not get a passport we would not be able to go
abroad, and she would go mad.

My own madness, or travel-induced eccentricity, left me gently.
I went through a phase of clearing out the trappings of my previous
life, throwing things away. My rented lock-up in West London,
with its piles of shirts, boxes of books and odd mementoes seemed
like the tomb of a being I barely recognised. There was little enough
there, but it was still too much: I wanted to slough it all off. The next
phase was a kind of manic environmental consciousness. I spent
days fiddling with the water and heating systems of Rebecca's
house, determined that we should live as lightly as possible. How
little you can manage with, in Africa, I kept thinking, and how much
we seem to need here. I found myself made twitchy and enervated
by the casual, unthinking way in which people shopped, their
habitual accumulation of objects. One day in a supermarket when I
overheard a little girl, who cannot have been more than nine, saying:
'Mummy, Mummy, stick to the list! We only came in here for three
things – stick to the list!'

Her mother was miles away, muttering prices to herself, her hands
reaching out for more 'bargains' as if of their own accord.

Having become accustomed to noting down and passing through, it
was difficult and unsettling to see these places, England and Wales, as
home; to see their people not as part of the continuity of peoples, of
varying tribes, stretching from the south of Africa to the north of
Europe, but as distinct, as special to me. I could not quite do it. Were
these really 'my' people? Was this 'my place'? I had lost faith in these
possessive distinctions.

There must be many who feel this way, but they are hard to meet.
I found one at a party in London, another ex-student of my old school,
Mack. Mack was wearing dusty-coloured trousers and a white shirt,
and he was tanned. A few years ago he drove a Land Rover from South
Africa back to Britain. He described the vehicle in loving detail, in a

sort of poem of jerry cans, tyres, reinforcements and customisations. It was the kind of conversation you relish in Africa.

'It needs a weld on the front right window post because it's flapping off.'

'Where is it?'

'In Portsmouth.'

I sucked my teeth. 'By the sea then?' (Salty air, corrosion – legend has it that Cape Town plates reduce the value of a vehicle, while a Johannesburg registration, indicating that it has been working in dry air, increases it.)

'Yeah but it's inside!'

'Ah, right. How much do you want for her then?'

'I'm asking for £4,000,' Mack said, reluctantly, as if the thought of selling such a trusted old friend pains him.

'That's a – democratic price . . .'

'No tax, no insurance, no one in Africa would insure it, shit all up the side, South African plates – but it does look fucking cool when you drive it through London.'

'Didn't anyone try to stop you?'

'Customs guy at Dover ran into the road with his arms up after I had passed, but I kept going . . .'

'Why are you selling her, then?'

'Because it's petrol. And because I want to design another one. The ultimate vehicle for Africa, basically.'

'You're going back then?'

'Yeah. It's all I think about. I just want to go back.'

When I heard that I could have hugged him. I was not so unbalanced or dislocated, after all, or at least, I was not alone in the dislocation. You do just want to go back.

The swallows raised their broods. I watched them everywhere, in Worcestershire, swooping over and partly into a pond outside Hereford; in Lichfield; in South Wales; in France again, briefly, and in Spain south of the Pyrenees, and on the Pennines, outside

Rochdale, flying in and out of a barn in the midst of light summer rain.

As the year turned, their young began to fly: I watched a mother schooling her short-tailed brood. First they went just a few feet, to a cherry tree near their nest where they perched, cheeping, waiting to be fed; the next day they went further; and on the third day they were flying and hunting.

When autumn came, the laws of migration were written in the movements of birds. The Canada geese which lived on the canal and the lakes began to move in V-shaped skeins, crying their wild call. On a walk on the moors we came across a flock of fieldfares, the northern thrushes, back from Norway to pillage the hawthorn berries and the yew trees. One day towards the middle of September the Rochdale swallows disappeared. I was writing when suddenly a large group, and some house martins, appeared in the sky above the beech tree beyond my window. They made their chattering, burbling cry, turned in the air, and then they were gone. On the 5th of October, walking out after rain in South Wales we found ourselves under a small cloud of swallows, twenty or so, all twittering and skidding in the air. They seemed to be following a black front of low pressure, matching the speed of the changing weather. They travelled with torn skies, white and azure, and dazzling, rain-washed sun. I have not seen them since except in dreams. Looking down the valley to the south-west, the way they must go, I saw it as part of a road, the swallow road, which leads to the other end of the world. They would fly south-west, for the Channel, and out across France, over the train tracks, where, somewhere, the conductor in crocodile cowboy boots will be thinking about his novel, and over Languedoc-Roussillon, where old Yves will see them and think the fair weather is ending, and over Spain, where the old lady in Zaragoza will be sorry to see them go, and know that the snows will soon come to the Pyrenees, and on down, over Denis and Judy in Madrid, and south to Andalusia, Gibraltar and the Straits, into North Africa, where Patrice might see them, if he is still there. And then my giri-giri friend will see them as they cross Niger, and Josephine, in Calabar, unless she has abandoned her brothers and moved to the city, and Pascale in Cameroon, and perhaps Bertrand, that laughing boy, will notice them as they cross Congo, following the great river to Aimée and Christelle and Dino in Brazzaville, and another few days will see them in Zambia, Botswana and Namibia in the Caprivi camp, where Christoph will note their arrival, and people who notice them will say the rains are on their way. Rick will be thrilled to see them again, he will set his nets in Bloemfontein in the hope of catching a bird with a ring that reveals the secrets of its journeying. And who would not want to go with them? Perhaps, if I had not gone into that Marrakech hotel at that moment, if I had chosen another chair, I might be going too, into all the freedom of the land and sky! It is a dizzying thought: to live again in the wonderful rhythm of change, in the unfamiliarity of every bed and the novelty of every morning; to compact all needs into a rucksack: to know again the completeness, the lightness, the self-sufficiency and irresponsibility of the traveller.

The sky was still hazed with dark grey rain. Would you really give it all up, I wondered – would you swap security and love and friendship for the swallow road? Would you exchange the urge to build something for the longing just to be? Would you do without the scents of autumn and the colours of winter, would you turn the world under your feet, rather than let the seasons turn you?

One day, perhaps, but not now. The swallows were flying so carelessly, with that joyful disregard of space and direction. I blessed them on their way, and suppressed the urge, like a summons, to go with them. They were twittering, calling to one another.

One of me does not make a what? I know you know it, but who said it first? It was Aristotle! (*Chelidon*, those Greeks called me, and held festivals to welcome me back from my travels. Pallas Athene once became me, and watched Odysseus slaughter Penelope's suitors from a perch on a beam.)

You all say it now: Italians, who call me *Rondine*; Spanish, who say *Golodrina* (in Aragon and Castile I am the bird that thaws the snows); French, to whom I am *Hirondelle* (you used to fear I could turn milk to blood in the udders of your cows!) and Danes, and Germans, and Dutch. I built the sky, if you believe the Austrians. I stole the fire from heaven and pulled the thorns from the head of Christ – how else did I get my blood-red cheek?

'*Zwaluw!*' say the Dutch, when they greet me. I am *Svala* in Swedish and my arrival must be a comfort there, for in my name is also 'to console'. They are all beautiful, my appellations, full of speed and turn and dash, like *svale*, in Danish, and the German *Schwalbe*.

It was a prehistoric Germanic speaker who named me Swalwon, before all these north names, a very long time ago. Naturally, the English try to put it simply, as the English love to do. The English call a spade a spade, but they call me 'European Swallow', 'Chimney Swallow', 'Barn Swallow', 'Our Swallow' and just plain old 'Swallow', most of the time.

I could go on (I am Tsi-kuk to the Cornish, Swallie in Lincolnshire and my Welsh name is Gwennol) but there is really only one division between you all – strange creatures! – and we might as well settle it now. The British and the French sum it up nicely: in Britain I do not make a summer; in France I do not make a spring.

Bibliography

Ayto, John, *Dictionary of Word Origins* (London, 1990)

Browning, Robert, *Poems*, selected by Douglas Dunn (London, 2004)

Butcher, Tim, *Blood River* (London, 2007)

Camus, Albert, *The Myth of Sisyphus and Other Essays* (Paris, 1942), tr. Justin O'Brien (New York, 1955)

Evans, Martin and Phillips, John, *Algeria: Anger of the Dispossessed* (Yale, 2007)

Horne, Alistair, *A Savage War of Peace: Algeria 1954–1962* (New York, 1977)

Knight, Cassie, *Brazzaville Charms* (London, 2007)

Priestley, Mary, *A Book of Birds* (London, 1937)

Shakespeare, William, *The Complete Works of Shakespeare*, ed. W. J. Craig (London, 1908)

Tate, Peter, *Swallows* (London, 1981)

Turner, Angela, *The Barn Swallow* (London, 2007)

Tydeman, W. E., *British Land Birds* (London, 1870)

Wernham, Chris *et al.*, *The Migration Atlas: Movements of the Birds of Britain and Ireland* (London, 2002)

White, Gilbert and Mabey, Richard, *The Natural History of Selborne* (reprinted London, 2006)

Acknowledgements

All thanks to Roger, Sue and Claire Paterson, for the days in Rasiguères, where I saw those five inspirational birds. Thank you Claire, especially, for your great kindness, help and encouragement.

Thank you, Angela Turner, and thank you, Rick Nuttall, for your time, trouble and priceless expertise.

In South Africa, thanks to Neville, Muriel and Clive Rubin; to Jonty and Anne Driver; and especially to my father, John Clare, for such a happy and informative week in Cape Town.

In London, thank you Alexander Clare, dear brother, for keeping me in touch with home – and the updates on where to avoid. Thank you, dear Mum, for your excellent advice and fearless encouragement.

Thank you, Judy and Denis and Jane Rafter, for being saviours in an hour of need. God bless you – on behalf of all of us who have knocked on your door.

All thanks to my treasured friends Julian May, Merlin Hughes, Anna Rose Hughes, Elizabeth Hughes, Sally Spurring, Gerard and Margaret Morgan-Grenville, Mohit Bakaya, Rob Ketteridge, Fliss Morgan, Chris Kenyon, Toby Lynas, Suzi Fogg, Tamsin Cooper and Lawrence Pollard for your love, kindness, spare rooms and wise counsel.

Many thanks to Robin Jenkins, Bushra Sultana, Ambar Rashid, Norddine Kamay and Rosie Ryan. I hope we all meet again in Essaouira.

In Rochdale, thank you Jenny and Emma Shooter, and especially

Robin Tetlow-Shooter for taking a strange immigrant into your lives, and for being so understanding of the peculiar habits of a writer. Thank you Jodi Trick, Jazz Powers, Esther Pryce, George 'Jud' Greenwood, Miria Griffiths and Janey Majid for your great kindess, and for making me feel so at home.

For the time to begin thinking about this book, I would like to thank the students and staff of Atlantic College, particularly Ken Corn and Dave Booker, for the residency, which was a joy.

For the assignment which turned into a path-finding mission, many thanks to Peter Browne and Sarah Spankie at *Condé Nast Traveller*. The opening lines of Patrick Kavanagh's 'On Raglan Road' are reprinted from *Collected Poems*, edited by Antoinette Quinn (Allen Lane, 2004), by kind permission of the Trustees of the Estate of the the late Katherine B. Kavanagh, through the Jonathan Williams Literary Agency.

For work on this book all thanks to Tif Loehnis, Alison Samuel, Rachel Cugnoni, Parisa Ebrahimi, Lisa Gooding, Stephen Parker and all at Chatto & Windus. Thank you Jeff Edwards, for your beautiful maps.

Thanks to Gill Coleridge, for your reading and suggestions, and thank you, above all, Rebecca Carter – superlative editor! Without your tremendous work and wonderful skill this book would have been a poorer thing indeed.

And to all those along the way, some of whom appear in these pages, and many of whom do not, who helped in so many ways, thank you. Thank you particularly, Mark Evans, Anna Reeve and Ndidi Nnoli-Edozien.

Finally, thank you, dearest RKS, for so much. This is for you.

Index